ADAM SPENCER'S NUMBER LAND

First published in Xoum by Brio Books in 2019

Brio Books Pty Ltd
PO Box Q324, QVB Post Office,
NSW 1230, Australia
briobooks.com.au

Text copyright © Adam Spencer 2019
Design, illustration and typeset copyright © Brio Books 2019

ISBN 9781925589924 (print)

This book is copyright. All rights reserved. Except under the conditions described in the Copyright Act 1968 (Aust) and subsequent amendments, and any exceptions permitted under the current statutory licence scheme administered by the Copyright Agency, no part of this publication may be reproduced, stored in a retrieval system, transmitted, broadcast or communicated in any form or by any means, optical, digital, electronic, scanning, mechanical, photocopying, recording or otherwise, without the prior written permission of both the copyright holders and the publisher.

The moral right of the author has been asserted.

Cover design, text design and typesetting by Xou Creative, xou.com.au
Author photograph by David Stefanoff
Additional cover photography from Adobe Stock
Endpapers courtesy of the brilliant Akiyoshi Kitaoka
Printed in China by Leo Paper Products Ltd.

To all my fellow
numerical nerds
... we *rock*!

Introduction

Welcome to *Numberland*!

And welcome to my personal take on some of the most fascinating, groundbreaking and jaw-dropping stories in the observable Universe. I say 'some' because as much as I would like, there's just not enough room in one book to cover all of the amazingness that has gone on over the past 12 months. *Numberland* runs to 416 pages — or '13 lots of 2 to the power of 5' — but we could have spent most of that dealing with just *one* of the incredible stories from 2018: how scientists took a photo of a black hole! Yes, you read that right: after decades of hard work, teams of scientists from around the world coordinated an array of 8 massive telescopes to capture an image of an object ... 55 million light years away. I'll just let that sink in for a second.

Okay, heart rate returning to normal? It's true (and the photo is incredible), but it wouldn't have been possible without numbers. Numbers and maths are the language we use to understand our world — and the worlds beyond. Whether it's calculating feats of incredible human achievement like say, running marathons, or saving lives using the Bee Gees' 'Stayin' Alive', or just making sure planes land

VII

safely, it's fair to say much of society would grind to a halt without the genius of maths, science and technology.

There's no getting around the fact that we humans are facing massive challenges. From climate change, to global conflict, to poverty and inequality right on our doorstep, we have work to do. Human advancement got us into this, but it can get us out, too.

I hate seeing a whale wash up on a beach with 40 kilograms of plastic in its stomach. But after decades of inaction, we now know we have a problem with the plastic not-so-fantastic and people everywhere are stepping up and making a difference. Continual scientific advancements mean we'll soon be able to generate energy more cleanly and cheaply, reuse or recycle our rubbish better, and clean up our oceans, waterways and cities. None of this would be possible without maths and science.

You've probably worked out the title of this book gives a cheeky nod to the one and only Lewis Carroll: writer, photographer and my personal favourite, mathematician. His fertile imagination has given generations of readers a lot of joy ... and probably just a little bit of confusion, too. Throughout *Numberland*, you'll meet some of Carroll's wacky anthropomorphic characters like the White Rabbit and the Dormouse. We'll also stop along the way and play a bit of ... chess! We'll hear from some of the greatest scientific brains throughout the ages. We'll learn about Ford circles. And curling. And honey bees as smart as 4-year-old kids. And Japanese poetry. And how you can win $6 million and maybe even free pizza!

Okay, heart rate way back up there again? Truly, I think there's something here for everyone.

This 6th book of mine as always has been a labour of love (and panic at the fast-approaching deadlines). But mostly love. It's an honour to be the University of Sydney's Mathematics and Science Ambassador and to work with some of the biggest brains around. Thanks to the fantastic puzzlemeister Sean Gardiner and the great Gareth White, maths whiz and all-round good guy. Now, if you think maths has some rabbit holes you should try the internet — I don't know where I'd be without Kate Hewson, researching goddess. My good mate the Surfing Scientist Reuben Meerman was a big help once again. As was David Harvey from UNSW. While I didn't understand everything he said to me, he's amazing. The brilliant young Bella Francis had a look at a few of the puzzles and snapped me back into line. Cheers, Bella. Thanks to the @qielves: John Cook, Alex Bellos, Steven Strogatz, et al. You make the numerical niches of the internet a wondrous place to visit. And the team at Brio Books — Rod, Jon, Roy and David — did a great job pulling the info together and making it look cool. What a ride, guys.

So sit back, take a deep breath and get ready to venture down the rabbit hole into *Numberland*.

As always let me know what you think, ask any questions, or point out any typographical gremlins you might find at adamspencer.com.au or message @adambspencer on Twitter. You little influencer you.

AS

Writing in a Number Wonderland

You may well have deduced that the title of this book is a cheeky homage to Lewis Carroll's classic novel *Alice's Adventures in Wonderland*.

Published in 1865, it's a weird and wonderful tale of a young girl — Alice — who falls through a rabbit hole into a fantastic world inhabited by all sorts of strange, anthropomorphic creatures ... and wordplay, nonsense, riddles and puzzles.

But I probably don't have to tell you that. Carroll's books routinely appear on 'best of' lists and have never been out of print. They've been translated into over 100 languages and, if you happen to own one of the surviving, original first editions, you'll be pleased to know it could fetch somewhere in the order of several million Aussie dollars.

In 1951, Walt Disney made an animated musical classic based on the book and, although it was considered a bit of a flop at the time of its cinema release, Disney showed it on TV as one of the first episodes of his *Disneyland* series. Suffice to say it did pretty well and would go on to earn a metric tonne of money for the company.

But back to Carroll's beloved (not to mention slightly wacky) anthropomorphic critters ...

Let's meet a few of them. And where better to kick things off than with the White Rabbit himself?

Lewis Carroll's real name was Charles Lutwidge Dodgson and, aside from being a bestselling writer, he kept himself busy as an Anglican deacon, a photographer and, my personal favourite ... a mathematician.

Alice's Adventures in Wonderland should not be confused with that other bestseller, Alex's Adventures in Numberland, by fellow mathematician Alex Bellos, to whom I also doff my cap.

YOU ARE HERE

1

The White Rabbit

I don't know about you, but I've never seen a rabbit wearing a waistcoat out in the wild.

Carroll's famous character is stylish but foppish — pompous sometimes, but grovelling at others. He's certainly inspirational, though, and has featured in dozens of films, TV shows, books and songs.

In Tim Burton's 2010 film, *Alice in Wonderland*, the White Rabbit works for the Red Queen, but is also a secret member of the Underland Underground Resistance, sent by the Hatter to search for Alice. In the film he's given the name Nivens McTwisp. He also appears in the video game adaptation, manipulating time and attacking using his watch.

In Australia, rabbits (white or otherwise) have a somewhat more dubious history. European rabbits (*Oryctolagus cuniculus*) were introduced in the 18th century with the First Fleet. They soon took over. Well, almost. By the early 1800s, their population was exploding — as were the methods of controlling them — but the Western Australian 'rabbit-proof' fence of 1907 was anything but, and the introduction of the myxoma virus in the 1950s did all sorts of collateral damage and only brought the population down to around 600 million. Admittedly, that was from an estimated *10 billion* in 1920.

But that's a drop in the proverbial for the venerable rabbit. The oldest fossilised 'rabbits' date back to about 53 Ma ago. What's that, you ask? Well, going back 53 Ma (mega-annum — or a million years) would take us to the Eocene epoch or Cenozoic era ... so quite some time ago. Bunnies are native to southwestern Europe (southern France and Spain) where they stayed, happily breeding and frolicking, for literally millions and millions of years. But at some point — either with or without human help — they hopped to it and colonised the planet.

The iconic illustration you see here of the White Rabbit is by the legendary Sir John Tenniel from the original edition.

Carroll had a challenging relationship with Tenniel, who ordered the first printing, some 2000 copies, be withdrawn as he wasn't happy with the print quality of his work.

Académie o' words

One of the most exclusive clubs in the world, the Académie Française, is a collection of 40 writers, academics and cultural guardians founded in 1634 with the purpose of protecting the French language.

They are charged with the near-sacred duty of updating the official French dictionary.

Their splendidly-detailed official robes cost in excess of $50,000 *a pop* and they carry very expensive ceremonial swords. But rather than battling invading hordes of Vikings headed for Paris or White Walkers advancing upon Westeros, their enemies are more ... grammatical.

Given that one of the Académie's members recently referred to the battle against 'brainless Globish' and the 'deadly snobbery of Anglo-American' you can assume they take the gig pretty seriously.

If you're looking for a job, there are currently 4 vacancies at the Académie. Word of warning, though — they've failed to agree on who should fill those spots for over two years now.

Maybe I should give them a call and try on a bit of *'Comment vous appelez-vous ... Je m'appelle The Spence'*?

There are 29 countries with French as their official language. In order of number of speakers, they are:

1. DR Congo
2. France
3. Canada
4. Madagascar
5. Cameroon
6. Ivory Coast
7. Niger
8. Burkina Faso
9. Mali
10. Senegal
11. Chad
12. Guinea
13. Rwanda
14. Belgium
15. Burundi
16. Benin
17. Haiti
18. Switzerland
19. Togo
20. Central African Republic
21. Congo
22. Gabon
23. Equatorial Guinea
24. Djibouti
25. Comoros
26. Luxembourg
27. Vanuatu
28. Seychelles
29. Monaco

Woman, 104, of Bristol, UK, arrested

At age 104 Anne Brokenbrow (by the way, *lit* surname, Anne) was arrested by police at her nursing home in Bristol in the UK.

The charge? 'Being a good citizen.'

Turns out Anne's lifelong ambition had been to be arrested. Reflecting on her law-abiding life till that point, she said, 'I am 104 and I have never been on the wrong side of the law.' In celebration of her 104th, the local law kindly obliged.

Never been on the wrong side of the law? Phftt. That's what Anne says. Perhaps she'd been an organised crime king-pin for the best part of 80 years and was thumbing her nose one final time at the authorities?

I've got my eye on you, Brokenbrow ;-)

PS —
Truly, here's to many more happy years on the right side of the law, Anne!

'Perfect numbers, like perfect men, are very rare.'

—René Descartes

For more on perfect numbers, (including what one actually is), see page 254!

A portrait of René Descartes after Frans Hals

🍪 + 🍪 + 🍪 = 30

🍌 + 🍌 + 🍪 = 14

🍌 + ⏰ + ⏰ = 8

⏰ + 🍌 + 🍌 × 🍪 = ?

This puzzle is bananas

B^{-a-n-a-n-a-s ...}

I have no idea where it originated, but I found it via the Twitter feed of — of all people — American actor and conservative posterboy @RealJamesWoods.

Have a go. As with many good puzzles, the answer doesn't involve any incredibly complicated trickery, but at the same time it's more difficult than it seems.

Good luck.

Check your answer where all good answers are found ... at the back of the book.

Once an obvious answer jumps out at you, perhaps you'd be best to take a deep breath and have another longer, slower look at it?

That's a big 10-4 from me, Rubber Duck...

Ah, breaker 1-9, this here's the Rubber Duck. You gotta copy on me, Pig Pen, c'mon? Ah, yeah, 10-4, Pig Pen, fer shure, fer shure. By golly, it's clean clear to Flag Town, c'mon!

These immortal lyrics come from CW McCall's 1976 hit 'Convoy' — a song that seeped so deep into popular culture it spawned a 1978 movie of the same name and saw a generation of kids my age running around the playground giving our friends a big '10-4'.

But what on Earth does it all mean? Where is Flag Town and who — or what — is the 'Rubber Duck'?

It's all part of the language developed by users of CB (Citizens Band) radio in the 1960s and 70s when speaking to each other, usually at night-time, usually from their bedrooms. It was sort of like the video gaming of the time ... but with fewer zombies.

CB radio was used by interstate truck drivers when communicating either for very practical reasons (warning of an upcoming accident), being a bit naughty (warning of an upcoming police speed trap), or just generally alleviating the boredom of driving through the night.

To translate the gibberish at the top of the page, it helps to break the communications into 3 categories. There are slang words, numerical codes and specific nicknames for locations along the highway. To make head or tail of it you need to know the following:

'Breaker' means 'I am about to break into this

channel and start communicating'. It is often followed by the number of the channel you are about to start talking on. In this case, channel 19 was very popular with interstate truckers. Here's a crash course on some of the other lingo.

'Got a copy' — 'can you hear me clearly?'

'Pig Pen' — a truck weighing station, also known as a chicken coop.

'10-4' — means 'acknowledged' or 'yes, I hear you'.

'Clean clear' — there are no police on the road.

Flag Town — Flagstaff, Arizona.

And Rubber Duck? That's the name of the big rig one trucker was driving.

So now it should be clear that the lyrics describe a trucking weigh station confirming to a trucker driving the Rubber Duck that there don't seem to be any police speed traps between his current location and Flagstaff, Arizona.

If you want to learn more about truck driver slang and visit a world where a 'ballet dancer' is a swaying antenna and an 'angry kangaroo' is a truck with defective headlights, where 'crackle-berries' are eggs and a 'pregnant rollerskate' is a Volkswagen Beetle, where Shaky Town is Los Angeles, California (the site of many shaky earthquakes) and taking a 10-100 is stopping to use the bathroom, well, I suggest you check out the book *CB Slanguage — the Official CB Radio Dictionary* by Lanie Dills.

Until then, this is Big Spence saying 10-7, I'm packing it in, good buddies. Wishing you 3s and 8s for the rest of your day in *Numberland*.

In Australia, the legal maximum length for any articulated vehicle (without a special permit and escort) is a stonking 53.5 metres.

Its maximum load may be up to 164 tonnes (gross), and it can have up to 4 trailers.

9

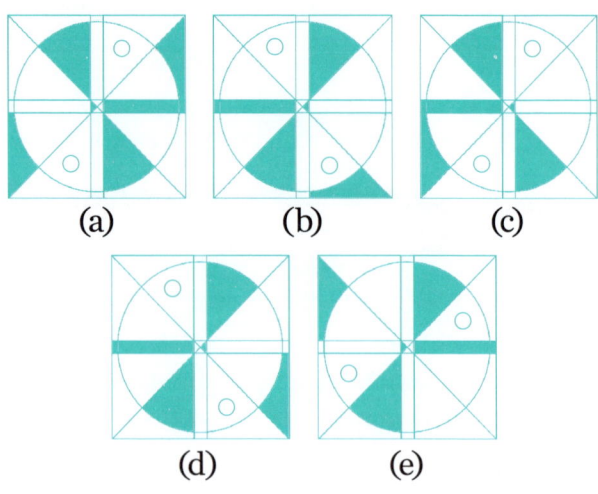

Kelly²

Science has long been fascinated by twins (and triplets) to study how we are shaped by hereditary factors and the world around us.

It's the age-old debate of 'nature versus nurture' and for the first time, scientists have been able to test their theories in the final frontier, space.

American astronaut Scott Kelly spent 340 days aboard the International Space Station (ISS) and, during that time, was subjected to a number of biological tests. Meanwhile, back on Earth, his identical twin, Mark, a retired astronaut, underwent the same tests.

Researchers noticed a variety of subtle changes in Scott's physiology as he floated above us.[*] Fascinatingly, many of his test results returned to match those of his twin brother's once he came back down to Earth in 2016, and he has remained in good health ever since.

It's not necessarily all good news, though. In one of the 10 evaluations, researchers detected a possible cause for concern. On several of the cognitive tests designed by

[*] The ISS orbits at an average of 408 kilometres above us. That's less than twice the distance of Tassie from the mainland of Australia.

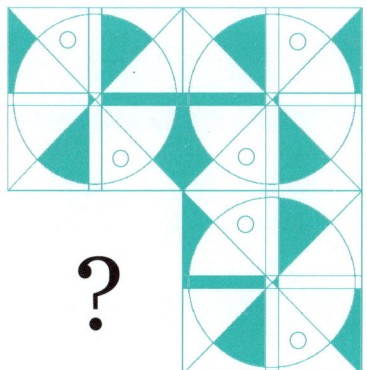

Mathias Basner of the University of Pennsylvania, Scott's accuracy dipped slightly in space ... and then fell even further on his return to *terra firma*.

Let's not forget that these tests were designed for astronauts — some of the most rigorously trained professionals anywhere on Earth ... or, ah, in space. So it's fair to say they're tough. Really tough. Basner points out that even he doesn't nail all the tests, all the time ... and he designed them. Oh, and the change in function was slight. Nevertheless, 'The decline and its protractedness are potential red flags,' he says.

One of the tests featured 3 geometric mosaics — the astronaut's task being to find the 4th mosaic which completes the set.

If you fancy giving it a go ... why, you're in luck. Strap yourself in and take a look above, my presumably Earthbound friend. The answer's right here on Earth — and at the back of the book.

ADAM SPENCER

My Very Excited Mother Just Served Us Nachos

No, it's not a declaration of my love for Mexican food (or more technically, American–Mexican food).

It's a mnemonic device (often a pattern of letters, which helps you remember something) to remind us of the order of … the planets from Mercury to Neptune.

Of course, my excited mother used to serve 'Us Nine Pizzas' … until the Pizza in that device, Pluto, was relegated …

Checkmate(!!)

Anyone who has read any of my previous books* will know that I love the game of chess as much as I love mathematics.

Perpetually frustrated by my limited ability, I love to read the plays of brilliant players and dream, even for a second, that I could one day dance around the 64 squares like the grandmasters of the world.

2019 saw two grandmasters (the highest ranking a chess player can attain — like black belts in martial arts) call it a day. The first was the great Vladimir Kramnik, who defeated the incredible Garry Kasparov and ruled as 'Classical Chess World Champion' from 2000—2006 before defeating Veselin Topolov, the then FIDE World Champion, therefore unifying the world title!

Yep, it's not just heavyweight boxers who unify world titles.

Big, bad Vlad, Special K, The Kraminator (all nicknames which I've just invented for him now) was one of the greats and in January 2019, aged 43, he said, 'I'm done'. Okay, more likely he said, 'Я закончила', which is pronounced '*Ya zakonchila*' … but you get my point.

Around the same time, one of the greatest Australian players of all time, Max Illingworth, also decided that he'd competed at the top level for as long as he could. I've been lucky enough to meet Max. He's a cracking guy and was Australia's 5th ever grandmaster.

So what better way to salute these two beasts of the board than by getting Max to give us his favourite ever Kramnik checkmate. How cool is this?

* What? You haven't purchased all 5 of my previous books?

Get your sweet little fingers to the nearest keyboard, type in adamspencer.com.au and knock yourself out! :-)

Let me set the scene. This match takes place, not at any old tournament, but at the Candidates Tournament: the showdown between 8 of the world's best players to decide who will go on to challenge for the world championship.

Kramnik is no longer world number one but, as he shows here, he still has the goods when it matters.

Against Levon Aronian (no slouch himself — ranked near the top in the world at the time) after 36 moves each this was the situation on the board.

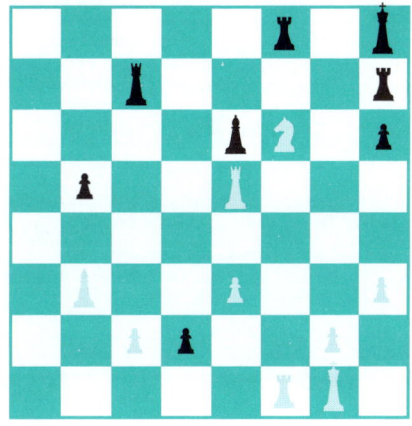

The squares are labelled with rows 1 to 8 and columns (we call them files) a to h. So the bottom left corner of the board is a1, the top left is a8, the top right h8, and so on.

Aronian, in a bit of trouble, had just moved the black queen to square c7. We chess nerds say 'Aronian played Qc7'.

Now, Kramnik can take the queen at c7 but Aronian will then capture the white queen with his rook which currently sits at h7. Without out the queens on the board, Aronian would take the heat out of the situation somewhat.

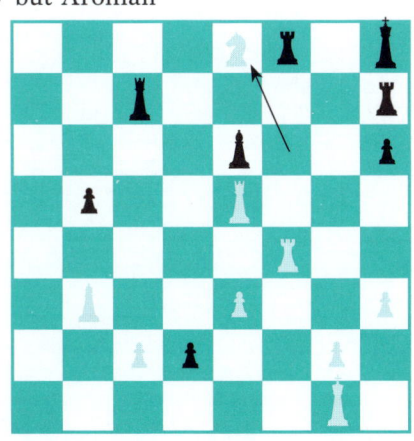

But not today, A-dog (again, my nickname, not accepted by the international chess-playing community as far as I know).

Kramnik unleashes and moves his knight (which some of you may know as his 'horsey') to e8.

15

I'll let Max take it from here. Suffice to say, the use of '!!' is chess-speak for 'Wow, that's an unbelievably awesome move'.

'37.Ne8+!! In my opinion, this is the most beautiful check-mating pattern that Vladimir Kramnik has played.

'With 37.Nf6-e8+, Kramnik unleashes a check to the black king on height, in full knowledge that black can take the white queen — 37... Qc7xe5.

'But after that, Kramnik can finish the game with 38.Rf1xf8+ — check to the black king! Black's only move is 38 ... Be6-g8, after which 39.Rf8xg8# is checkmate — the rook is defended by the white bishop on b3.

'Note that other ways for black to get out of check allow white, to, sooner or later, take the black queen, or even give checkmate in several moves.'

Yep, good choice, Max. Vlad, even by your standards, that's an !! move.

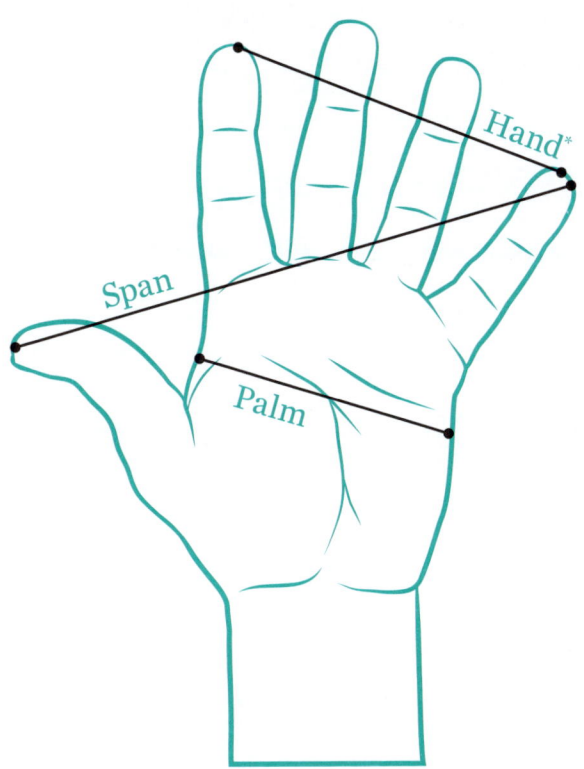

Attention, span

A 'span' is traditionally the distance on an adult male hand from the tip of the thumb to the tip of the pinky finger.

It is officially 9 inches (or 22.86 centimetres for everyone enjoying the metric system).

In the Slavic languages, such as Russian, Ukrainian, Belorussian, Polish, Czech, Slovak, Slovenian, Bosnian/Croatian/Serbian, Macedonian, Bulgarian ... and so forth, they speak of a *pyad*. That's the distance from the tip of the thumb to the index finger (or forefinger or pointer finger) — or 7 inches (17.78 centimetres).

Try slipping 'damn — missed it by a *pyad*!' into conversation next chance you get.

* Note the 'hand' measurement is actually from forefinger to pinky with fingers closed together. A 'span', on the other, ah, hand ...

Here are two more BONUS archaic hand units of measurement. Phwoar! Thanks, Adam!

1. Hand = 4 inches (10.16 centimetres)
2. Palm = 3 inches (7.62 centimetres)

17

Ants comprise 15—25% of the Earth's terrestrial animal biomass

Which is to say that the 20,000 or so known species of ants account for up to a quarter of the animal biomass on the Earth, according to entomologist Ted R. Schultz.

My own research suggests that, at any given picnic, ants will account for a significantly greater percentage of the biomass on any given slice of pavlova.

Six of the best

Sir Martin Rees is a pretty cool dude. For a number of reasons.

Firstly, he is a world-class scientist who specialises in cosmology and astrophysics.

Secondly, he is the British Astronomer Royal.

The Astronomer Royal is a senior post in the United Kingdom's Royal Household that dates all the way back to 1675. Sir Martin is the 15th man to hold the position. It's an unfortunate sign of even these slightly more enlightened times that you don't really need to ask how many women have held the position ... do you?

While the position is largely honorary these days, the Astronomer Royal does remain available to advise the Royal Family on astronomical and related scientific matters.

So, strictly speaking, if Harry and Meghan were arguing over the number of stars in the Milky Way, it would be Sir Martin's job to intervene and suggest, sensitively, 'Well, the best estimates are in the range of 100 to 400 billion, your highnesses'.

The third really cool thing about Sir Martin — and what earns him a spot in *Numberland* — is a book he wrote back in 1999 entitled *Just Six Numbers*.

In this cracking read, Sir Martin describes 6 numbers that are especially significant in underpinning the fundamental laws that govern our Universe. In his words, 'Two of

them relate to the basic forces, two fix the size and overall "texture" of our Universe and determine whether it will continue forever, and two more fix the properties of space itself'.

What are the 6 numbers, you ask?

$N \approx$ 1,000,000,000,000,000,000,000,000,000,000,000,000

N is the strength of the electrical forces that hold atoms together divided by the strength of gravity between them. It represents how much 'weaker' gravity is than the forces going on inside atoms. If N had a few less zeroes in it gravity would be too strong and nothing of significant size could exist without being crushed.

$\varepsilon \approx 0.007$

The nucleus of a helium atom contains two protons and two neutrons. But the nucleus weighs 99.3% of what two protons and two neutrons weigh. The 0.007 of the mass that is 'missing' is given off as energy, mainly heat, when two hydrogen atoms fuse. So $\varepsilon \approx 0.007$ is the 'efficiency' at which hydrogen atoms fuse into helium atoms in stars like our Sun and determines how long stars live. Stars are crucial for the formation of heavier elements and life itself, so well done, ε.

Ω ≈ ? (We're not sure, but it's really close to 1)

This is a measure of the mass density of our Universe. If Ω > 1 our Universe would end by collapsing back onto itself in a 'big crunch'. If Ω < 1, our Universe will keep expanding forever. At the moment we just don't know, making this one of the most exciting questions in modern physics.

Λ ≈ 0.7

This is an indicator of the rate of expansion of the Universe. Thankfully this number is very small. If Λ was too large the Universe would be expanding too rapidly for stars and galaxies to have formed.

Q ≈ 1/100,000

This governs the structure of the stars, galaxies, clusters of galaxies and even superclusters. The fact that our Universe is neither rampagingly violent, nor bland to the point of nothingness, is due to our good friend Q.

D = 3

This refers to the number of dimensions of space in our Universe.

What's really incredible is how finely-tuned these 6 dials are to support everything that is going on.

For example, Sir Martin points out that if ε were 0.006 or 0.008 instead of 0.007, 'we could not exist'.

And wouldn't that be an (Astronomer) Royal pain in the bum!

8-0 — it's a snowman!

Curling is a fascinating game.

It's sort of like lawn bowls — but it's on ice instead of grass, and people use brooms, and … look, it's probably best you watch a game or two online to get the idea. Or hang around till the next Winter Olympics.

Anyway, in a single 'end', each team throws 8 'rocks' and the team that has the closest rock to the 'button' (the centre) of the concentric circles (called the 'house') scores a point for each rock inside the house that is closer to the button than their opponent's most central rock.

You with me?

In the best case scenario, if you get all 8 of your rocks inside the house — I know, it sounds like commercial radio! — and they are all closer to the button than any of your opponent's, you score 8 points for the end. It's the most you can score on an end and is an extremely rare occurrence. And, even cooler, it's called a snowman.

Want more? Okay, curling teams are made of 4 people (each player throws two rocks). Two teams play against each other on one 'sheet' (lane) of ice.

The Dormouse

Lewis Carroll's Dormouse is known for falling asleep and serving as a cushion for the March hare and the Hatter. He also joins the Tea Party's fun and games, telling bizarre tales about 3 little girls who live in a treacle well. Sweet!

A curious and curiouser character, to be sure. Such a trial, dear Sir, I'll stop right there!

Did you know the rodent (order *Rodentia*) comprises over 2000 species and makes up the largest group of mammals on the planet? They live everywhere except for Antarctica (too cold) and, by 2050, New Zealand (thanks to their ambitious 'world-first' project to make the nation predator-free), and include porcupines, squirrels and gophers in the extended 'family'. One feature they all share are their incisor teeth ... which never stop growing(!)

The house mouse (genus *Mus*) is typically a little critter (under 12 centimetres long) native to Central Asia. If you're a human, you'd probably say mice have super cute teeny pointed snouts and small rounded ears. If you're an elephant, however, you may see them as terrifying and unpredictable beasties. Of course, there's absolutely no truth to the elephantine myth ... it's pure folklore.

Mice are popular lab animals because, well, they're small and cheap to rear. And they breed like ... rabbits. They're pretty mild-mannered too, until you provoke them and they unleash their inner fury. PETA claim more than 100 million lab mice and rats are killed each year in the US alone. A horrible stat, but then countless medical breakthroughs (organ transplants, joint replacements, blood transfusions and chemotherapy to name just a few) came about because of mice sacrifice.

What's going on in the illustration opposite?

Spoiler alert ... 'The March Hare and the Hatter put the Dormouse's head in a teapot'.

Illustration by John Tenniel

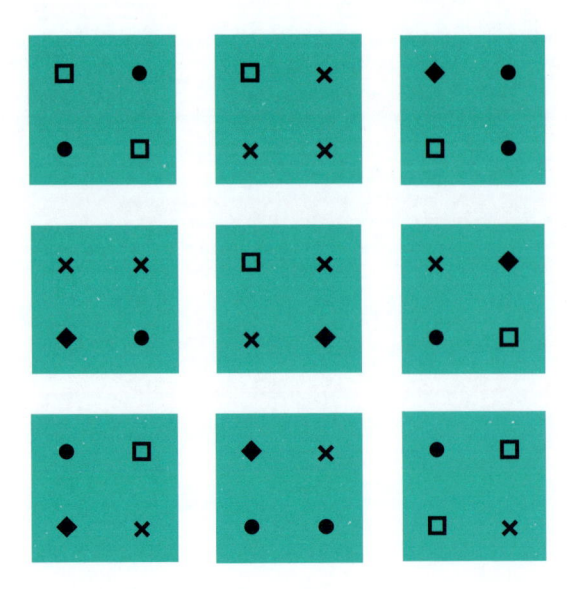

On the tiles in *Numberland*

S pare a thought for tilers in *Numberland*.

Not content with merely following the latest decorating fad for this bathroom reno, they have to contend with the following instructions.

Place the 9 tiles into a 3 × 3 grid so that adjacent tiles show the same symbol along their touching edges.

The tiles here are all up the right way — you don't need to rotate them. As another hint for your early tiling days, the top left hand tile is already in the correct position. But any or all of the other eight might need to be moved around.

Step to it!

As always, you can head over to my website to download a cut-uppable copy of this puzzle.

Point your browser to adamspencer.com.au

26

The *worst-*ever result in a chess simul occurred in Moscow's Soviet Pioneer Palace in 1951

A simul, or 'simultaneous' match is where one chess player plays multiple opponents all on different boards, at the same time. They are usually excellent opportunities for club players to play against masters — where the amateur usually loses ... eating humble pie at the master's display of intellect and prowess. But not always.

Back in 1951, the great British player, Robert Graham Wade, fronted up against 30 local school-kids aged 14 and under. After around 7 hours playing, International Master Wade made 10 draws, losing the other 20 games! Let's just say the kids were loving themselves sick*.

*There is a far more successful and historic simul coming at you ... in just a few pages!

On your marks ...

The Comrades Marathon is an ultramarathon footrace established in 1921 which takes place annually in the KwaZulu-Natal Province of South Africa.

While the technical length of a marathon is 42.195 kilometres (as I'm sure you'll know ...) runners in the Comrades travel a distance of roughly 87 kilometres between the cities of Durban and Pietermaritzburg.

It proudly claims to be the world's largest and oldest ultramarathon race but, before you go and grab your Nikes, you'll need to be 'ASA technically compliant' to compete ... which basically means you'll need to confirm you're capable of making it to the finish line in something resembling one piece.

The qualifying table lists minimum distances — and their corresponding times — as anything between 42.2 kilometres (marathon-length; in no more than 4:49:59) and 100 kilometres (no greater than 13:29:59, thank you very much).

And if you want to win it? Well, ballpark time would need to be somewhere in the 5 to 6 hour mark.

Incidentally, the Comrades is one of the roughly 1200 ultramarathon races this year ...

... In which I will not compete!

'The science of operations, as derived from mathematics more especially, is a science of itself, and has its own abstract truth and value.'

—Ada Lovelace

Portrait of Ada Lovelace by Margaret Sarah Carpenter

The Mirabal sisters

During the late 1950s reign of the Dominican Republic dictator Rafael Trujillo (El Jefe) 4 incredible women stood up against him.

The Mirabal sisters — Patria, Minerva, Maria Teresa and Dedé — engaged in secret activities aimed at mobilising people against Trujillo's brutal rule.

They were arrested but this did not deter them.

Tragically, 3 of the sisters, Patria, Minerva and Maria Teresa, were assassinated by Trujillo's henchmen on 25 November 1960. Trujillo himself would be assassinated a year later.

The sisters' inspirational struggle saw them described in the *New York Times* as 'symbols of both popular and feminist resistance' and in 1999 the United Nations honoured them by dedicating 25 November as the International Day for the Elimination of Violence against Women.

Each legal move in chess opens up roughly 38 *new* moves

Though estimates vary (32 new moves is a more conservative guess), if we assume that the average game of chess consists of 40 or so moves per player, that means the total number of games possible could be something like 38^{80} — or over 2×10^{126}. That's a 2 ... with 126 zeroes after it.

That should keep you entertained for a while, given that there are 'only' 10^{86} elementary particles of matter in the observable Universe.

ADAM SPENCER

'Everyone, at any age, has talents that aren't fully developed — even those who reach the top of their profession.'

—Garry Kasparov

Kasparov, age 11. Copyright 2007, S.M.S.I., Inc. Owen Williams, The Kasparov Agency.

Rage against the machines

June 6, 1985, Hamburg.

Garry Kasparov, who would later that year become World Champion and is still considered by many to be the greatest player ever, faced off against the world's best 32 chess computers ... all at the same time.

As we learned a couple of pages ago, this form of chess is called a 'simul', short for simultaneous play.

Kasparov was such a chess personality, several of the computers he played against were actually branded 'Kasparov models'. In the end, humanity — or more specifically, Garry — won 32-0. In your face, robots.

But it did get close for a moment.

According to Kasparov, one opponent, Turbostar 432, had a chance to win but moved the queen to the wrong square on the 37th move of their clash. The soon-to-be World Champ had to be particularly careful against the Turbostar, precisely because it was a Kasparov-themed opponent. 'At one point I realised that I was drifting into trouble ... if this machine scored a win or even a draw, people would be quick to say that I had thrown the game to get PR for the company, so I had to intensify my efforts,' said Kasparov in Diego Rasskin-Gutman's *Chess Metaphors: Artificial Intelligence and the Human Mind*.

'Eventually, I found a way to trick the machine with a sacrifice it should have refused. From the human perspective, or at least from my perspective, those were the good old days of man versus machine chess.'*

---

* Of course, 'the good old days' couldn't last forever.

Just 12 years later, on 11 May 1997, the IBM Chess Computer Deep Blue defeated Kasparov $3\frac{1}{2}$ - $2\frac{1}{2}$ in a best-of-6 game contest.

In just over a decade, we had gone from the World Champion walloping the best 32 chess computers in the world, *at once*, to falling just short against a single machine.

These days I have an app on my phone that would crush any human into chess-playing dust.

It cost me $9.95 bout 5 years ago!

n

Did you know that 'for every number n, you can find a span of n consecutive numbers containing no primes'?

Well, the good people at @AlgebraFact clearly knew this because they tweeted it as their daily truth-bomb on 5 March 2019.

But what does it mean?

Let's recall a bit about prime numbers. The number 6 is *not* prime because we can write it as the product of 2 and 3; namely 6 = 2 × 3. We say that 2 and 3 are 'factors' of 6. The number 7 *is* prime because its only factors are 1 and itself. While we can write 7 = 1 × 7, there is no way to break 7 into the product of smaller factors.

Every whole number greater than 1 is either prime or not prime. We sometimes call 'not prime' numbers 'composite'.

It should be obvious that all even numbers greater than 2 are composite, because they are divisible by 2.

So having a look at the numbers in the 40s — why not the 40s hey, they're just as cool as any other numbers and rarely get the love! — we see that 40 is composite (40 = 2 × 20 = 4 × 10 = 8 × 5) as are 42, 44, 46 and 48 (all even), 41 is prime, 43 is prime, 45 is composite (45 = 3 × 15 = 9 × 5), 47 is prime and 49 being 7^2 is composite.

Now, just before the 40s we have 37 (prime), 38 and 39 = 3 × 13 (composite) and just after the 40s we encounter 50 (even), 51 = 3 × 17 and 52 (even) which are all composite before we run slap bang into the prime 53.

So starting from 37 and counting up to 53, in terms of primes and composites we have:

P (37) C C C P C P C C C P C C C C C P (53)

... during which the longest stretch of composite (or non-prime) numbers we encountered was the span of 5 numbers from 48 to 52.

QUESTION: Before you check out the table below, if we take the integers 1 to 100, there is only one span of composites longer than 5. Can you find it? Take a moment to try this before you go any further.

If you look at just the first 100 counting numbers, you could be excused for thinking that the spans of composites will only ever be 1, 3 or 5 long with the occasional 7 thrown in.

But the claim made by @AlgebraFact is that no matter how big a number n you nominate, there has to be a span of n composites out there, somewhere in the counting numbers. There must be a span of length 11, of length 100, of length a googol (1 followed by 100 zeroes). But we are working up a sweat trying to find a span of even length 7.

What gives?

It turns out that when we start off counting from one, early on, we meet up with primes all over the place. But as

1	2	3	4	5	6	7	8	9	10
11	12	13	14	15	16	17	18	19	20
21	22	23	24	25	26	27	28	29	30
31	32	33	34	35	36	37	38	39	40
41	42	43	44	45	46	47	48	49	50
51	52	53	54	55	56	57	58	59	60
61	62	63	64	65	66	67	68	69	70
71	72	73	74	75	76	77	78	79	80
81	82	83	84	85	86	87	88	89	90
91	92	93	94	95	96	97	98	99	100

we get further into the counting numbers, the primes thin out a bit.

There are 25 primes from 1 to 100 or 25% of the first 100 counting numbers are prime.

But as we look at the first 1000 or 10,000 counting numbers, check out what happens:

x	Number of primes less than x	
10	4	40%
10^2	25	25%
10^3	168	16.8%
10^4	1 229	12.29%
10^5	9 592	9.59%
10^6	78 498	7.85%
10^7	664 579	6.65%
10^8	5 761 455	5.76%
10^9	50 847 534	5.08%
10^{10}	455 052 511	4.55%
10^{11}	4 118 054 813	4.12%
10^{12}	37 607 912 018	3.76%
10^{13}	346 065 536 839	3.46%
10^{14}	3 204 941 750 802	3.20%
10^{15}	29 844 570 422 669	2.98%
10^{16}	279 238 341 033 925	2.79%
10^{17}	2 623 557 157 654 233	2.62%
10^{18}	24 739 954 287 740 860	2.47%
10^{19}	234 057 667 276 344 000	2.34%
10^{20}	2 220 819 602 560 910 000	2.22%
10^{21}	21 127 269 486 018 700 000	2.11%
10^{22}	201 467 286 689 315 000 000	2.01%
10^{23}	1925320391 606 800 000 000	1.93%
10^{24}	18 435 599 767 349 200 000 000	1.84%
10^{25}	176 846 309 399 143 000 000 000	1.77%

So as we stretch out into larger and larger numbers, the primes are a lot rarer than they are in the realm of smaller numbers in which we mostly move.

As the primes thin out, it should make sense that we might start to encounter longer and longer runs of composite numbers.

In case you're wondering (and who *am* I kidding) the first run of over 500 consecutive composites is the run of 513 numbers from 304599508538 to 304599509050. But finding a run of 500, or 5,000 or 100,000 composites doesn't prove the claim that started all this — that you can find a run of *any* length you want. Well maybe this will work for you: for any positive integer n, the sequence of n consecutive integers $(n + 1)! + 2, (n + 1)! + 3, ..., (n + 1)! + (n + 1)$ contains no primes.

Let's check out why this is the case.

We know from the definition, that:

$$(n + 1)! = 2 \times 3 \times ... \times n \times (n + 1), \text{ for any integer } n.$$

It should be clear that $(n + 1)! + 2$ is divisible by 2, because $(n + 1)!$ has a factor of 2 and we've just added 2 to it.

So $(n + 1)! + 2$ must be composite.

But by the same logic, $(n + 1)! + 3$ must be divisible by 3 and therefore composite. $(n + 1)! + 4$ must be divisble by 4 and therefoe composite ... and so on.

So the string of n numbers $(n + 1)! + 2, (n + 1)! + 3, (n + 1)! + 4, ... (n + 1)! + n + 1$ must *all* be composite.

Therefore, no matter how long you want your string of n composiites to be, there is one out there at least that long starting from $(n + 1)! + 2$. It might not be the first one hidden in the counting numbers but it is certainly there ... for *any* length n you wish to find.

Well done, @AlgebraFact.

On a separate note, the mathematiics around the distribution of prime numbers is some of the deepest and most famous work ever done. If you'd really like to have your world rocked, look up the 'Green-Tao theorem'.

ADAM SPENCER

→	5	+	6	×	4
	+	5	×	5	×
	8	+	8	+	7
	×	7	−	8	=
	8	×	9	=	28

Snakes on a plane

I n the immortal words of Samuel L. Jackson ...

Everybody strap in. I'm about to open some windows*.

In the following grids, you can make paths from the top-left corner to the bottom-right by moving between adjacent squares.

For each grid, see if you can find 3 different paths that all trace out correct equations.

Order of operations applies! (Remember, per our handy *Numberland* mnemonic on page 270: Pandas Eat Milk Duds And Skittles ...)

I've provided an example here to kick things off, but after that, you're on your own.

See if you can make the firetruckin' snakes out of our firetruckin' grid.

And if you can solve these, well done. They are among the most fiendish puzzles I've asked in my many books!

*If you're unfamiliar with this now-classic movie quote, by all means head on over to YouTube.

But brace yourself for some, ah, Samuel L. Jackson-like language ...

YOU ARE HERE
38

NUMBER LAND

→	5	+	6	×	4
	+	5	×	5	×
	8	+	8	+	7
	×	7	−	8	=
	8	×	9	=	28

→	5	+	6	×	4
	+	5	×	5	×
	8	+	8	+	7
	×	7	−	8	=
	8	×	9	=	28

→	5	+	6	×	4
	+	5	×	5	×
	8	+	8	+	7
	×	7	−	8	=
	8	×	9	=	28

Got the drift?

    Okay, now it's your turn …

→	3	+	5	−	1
	−	4	×	6	−
	1	×	3	×	6
	×	5	×	9	=
	1	×	2	=	1

YOU ARE HERE

39

Oceania, including Australia, had 28,634,278 internet users as at 9 May 2019

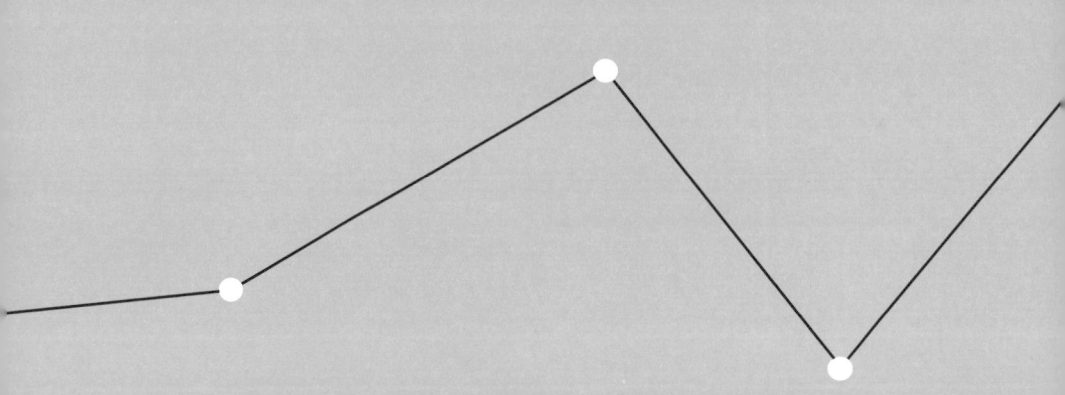

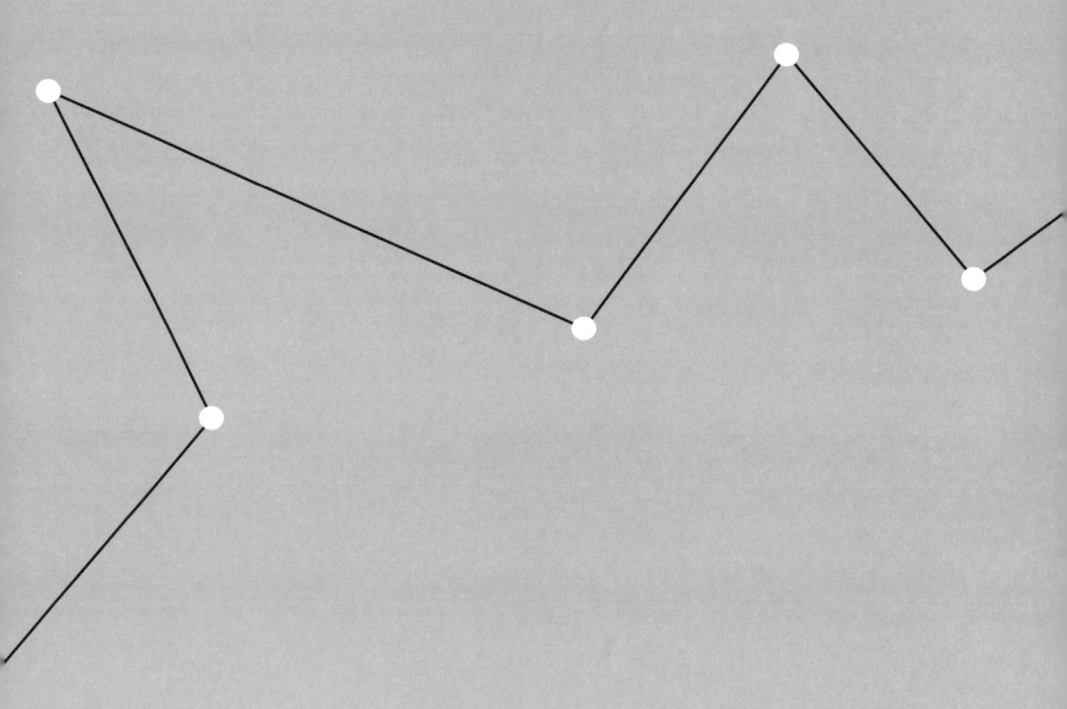

That accounted for roughly 0.7% of the world's internet users, according to internetworldstats.com.

For reference, the region was determined to have an overall population of 41,839,201, accounting for 0.5% of the world's population and meaning a 'penetration rate' of 68.4%.

Sublime prime

You might recall from my *Big Book of Numbers* that, since 31 = 2^5 − 1, it is the third Mersenne prime.

Here's a refresher from that book.* Almost all of the largest prime numbers ever discovered are Mersenne primes (primes equal to 2^p − 1, where *p* is prime).

The 48th January 2013 by Dr Curtis Cooper, is:

$$2^{57,885,161} - 1$$

It has a whopping 17,425,170 digits and if typed out in the same font as the 7 *Harry Potter* novels would run half that length again — an incredible 5000 or so pages.

To find out more about these massive primes, search the web for the TED talk given by a dashingly handsome Australian number-loving comedian ... what was his name again ... I've got it lying around somewhere ... that's right ... it was ME!

Initially, the Mersenne primes occur fairly regularly:

2^2 − 1 = 3 is the first Mersenne prime, written as M_1;
2^3 − 1 = 7 which is M_2; 2^5 − 1 = 31 = M_3; 2^7 − 1 = 127 = M_4

... but just when you're beginning to think, 'I'm onto something here: for any prime number *p*, the number 2^p − 1 is also prime! Wow ... they'll call this the (insert your name here) theorem ... and I'll be famous!' ... you realise that:

2^{11} − 1 = 2047 = 23 × 89 ... which is *not* prime.
So 2^p − 1 is not prime for every prime *p*.

One of the beautiful things about maths is that it never stops advancing. The text above from 2014 about the largest prime number we had yet discovered was one of my favourite chapters of my first book.

* (Cue the harp sounds).

** (Cue the record scratch sound).

YOU ARE HERE!

42

It particularly resonated with me because it was the topic of my 2013 TED talk which, while I joke about it, is actually a pretty good watch if you've got a spare 18 minutes.

Now, every time a new Mersenne monster is found, my nerdy heart skips a little beat. That's why I'm so excited to update this entry and present it to you here.

I especially love that in a decade we will look back and have a good old chuckle that we thought *this* was a massive prime! Sure, at almost 25 million digits it is around double the length of the entire *Harry Potter* series.** But in the infinity of primes that lay out there … it's still just a bubba!

$$2^{2203} - 1$$
(October 1952, 664 digits long)

$$2^{11,213} - 1$$
(June 1963, 3376 digits long)

$$2^{21,701} - 1$$
(October 1978, 6533 digits long)

$$2^{216,091} - 1$$
(September 1985, 65,050 digits long)

$$2^{1,398,269} - 1$$
(November 1996, 420,921 digits long)

$$2^{43,112,609} - 1$$
(August 2008, 12,978,189 digits long)

$$2^{57,885,161} - 1$$
(January 2013, 17,425,170 digits long)

$$2^{82,589,933} - 1$$
(January 2019, 24,862,048 digits long!)

In case you're out looking for the largest yet-known prime, it starts a little something like:

148894457420413255478064584723
979166030262739927953241852712
425213239361064475310309971132
033717475283440142358756005197
183265856491842931959708229506
433434510973136992053423106411
059526476787674681933221178184
754771079862112265347927886299
212447235816979464642467372269
115661546889834987857788089927
633363565129754335286257452179
554111565785480230295382592318
040461918808066672007922224457
059309881538873940476999622792
719431939650771206572696591287
891780444893214525405268925811
669721358726058130396831449510
439814585421184420014843770161
642903895817082977059418889948
932701608127972741434818590807
599648655190062672294171521513
452828119103082446114401235115
568521967470388265793076255199
415833523853151542818458668825
535895472102988098477808837016
635141972524013277223153442722
747181306147625815374655866269
838102926072292274274159167780
409861935722047159366119319961
718058420541094365289984777531
826224519087060254159129057555
034019195752086990922805950586
234834234333902221578051754478
315206811414437205217972195325
092355278128460175429150099729
338701354569529879819532035048
795142078820863181303301447893
004993880945511123110175951270
751799108933054789684767388453
528956294865410389965240118794
202304359822718727319453928622
404354611551920664726615294736
664913439805179135241335847547
822227043388948929318395674897
318657027251644004792296222422
578898843573334941233981420990
503634531584014992359905105152
214472444022706258956755283134
259132359157427762069987246226
943677702099990555277196120271
974172256327014788875746679124
936671482047022971609066596465
712569235389178681061603854163
840520001622519567396714687649
494867746446903248286792594583
481446378168583826791675523408
671580058913077635720983396099
051753835595984597239296163028
719794883298480139380579804561
405758687738625458851970408170
333412776131402799624362356692
427713182265102353818596183931
065827456075052455166799215412
831030874556394034080539903165
972717244664991017846982996394
530760748079971380078450714607
897701462244102623923688549199
403622390511150372397658844727
492526195023383876899707216045
726690363728261075630868392922
589921984272926571696420949013
985487230937139609819688831018
277160960191570936096818063132
646300158015504854717539777024
516996167655901390832104325224
732353244066633538619838088794
165029562282047651893598037393
666631169200343587494296098854
795275523568553751266232459443
520299164600539999992380182262
310208184982948769365533292206
326471982948111697436839933559
987974930955023234966439599006
640854230611034282445894395210
549716035982251112894912920223
973271085559610694387817701316
441685157857081050023685868390
959945692549045316738691873685
286171869477177485400278433762
007956776070022608536807517839
506668702417880021319420490830
544260824154115722969256967502
756610140232426411929630091908
819222213572079714547598399790
153529120046061849130603031714 ...

And just because I can tell you … it ends in a 1.

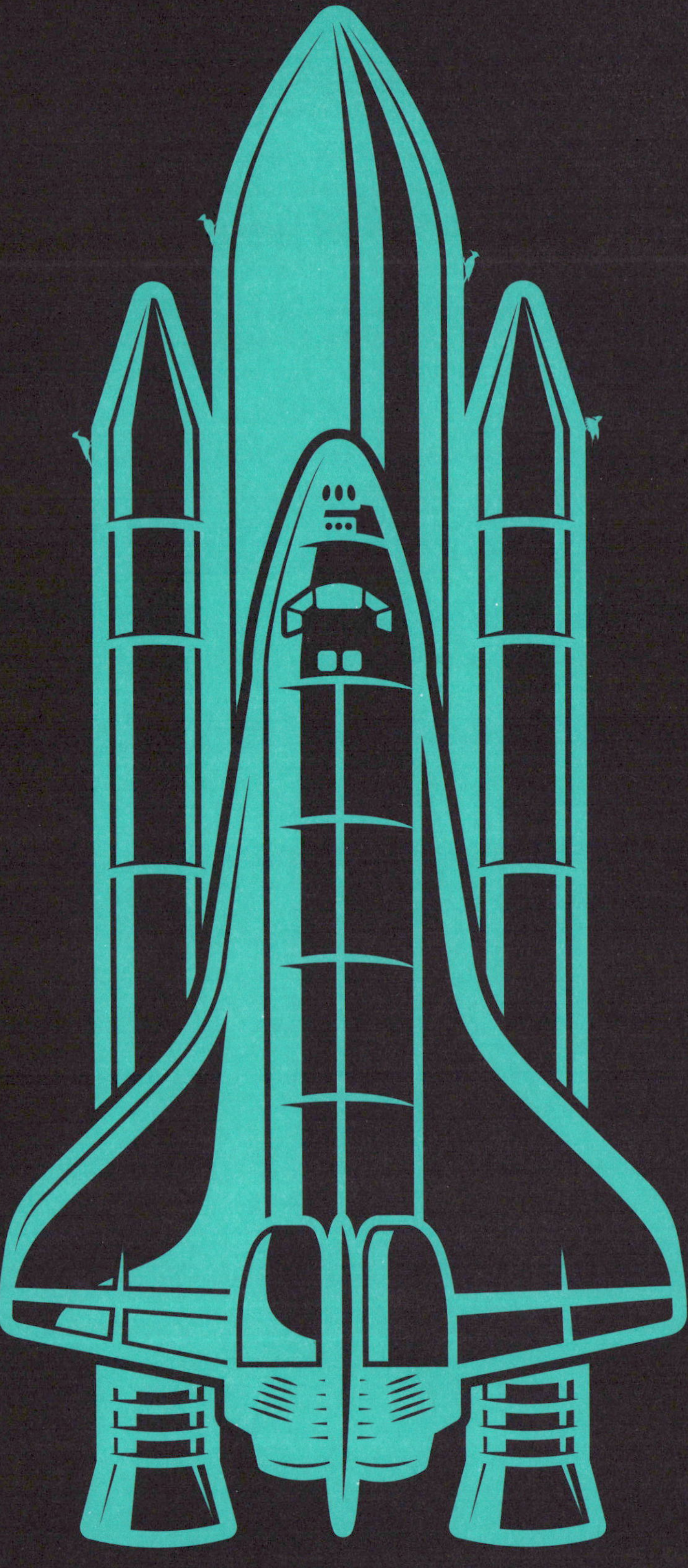

The NASA space shuttle *Discovery*'s 1995 launch was delayed by (at least) 71 holes in its foam insulation

The half- to 4-inch deep holes were made by a group of (presumably disappointed) woodpeckers, thinking they'd found the mother of all trees.

To ensure future launches weren't similarly disrupted, NASA deployed … decoy owls.

Sadly, some years later, problems with this same insulation proved fatal for the *Columbia* and its crew.

A whole lotta yotta

Computerised data is growing.

And I mean *growing*. In a decade or so, it's estimated the total amount of data worldwide will exceed 1 yottabyte (or, one septillion [10^{24}, or a 1 followed by 24 zeroes] bytes) ... but if we don't expand our list of prefixes, we'll have no way of describing it.

Begin the name game! Head of metrology at the UK's National Physical Laboratory, Dr Richard Brown, fears we may be headed for a 'Boaty McBoatface'* scenario when it comes to weights and measures unless we get a-namin' now. And while the inner kid in me says, 'Cool, bra', my inner scientist says, 'Nah way.'

So what do we choose? Prefixes like 'bronto' or 'hella' are pretty popular, but what about 'tyranno', 'stego' and 'colosso'? Dr Brown likes a laff — don't we all — but remains unconvinced by such prehistoric naming notions. He suggests we go slightly less old-school — 'ronna' and 'quecca' for octillion (27 zeroes) and 'nonillion' (30 zeroes), along with 'ronto' and 'quecto' for their fractional counterparts, octillionth and nonillionth. Good old fashioned Latin and Greek words ... in other words.

Well, they're not *actually* Latin and Greek. Ronna and quecca are deliberate mispronunciations of 'ennea' (Greek for 'nine') and 'deka' (Greek for 'ten'). This tradition started because the prefix 'tera' (from the Greek for 'monster') looked like a typo of 'tetra' (Greek for 'four'). Ah, language.

So perhaps Boaty McBoatface-byte stands a chance, after all?

* 'Boaty McBoatface' is the name of the lead sub carried on the research vessel RRS *Sir David Attenborough* owned by the Natural Environment Research Council (NERC) and operated by the British Antarctic Survey (BAS). It was named following an online public poll ... a public poll which, it's fair to say, went in a direction that the organisers were *not* expecting.

X marks the spot

NASA's *X-34* space planes were developed in the late 1990s out of a desire to make it cheaper and safer to get people and cargo into space.

The *X-34* was to be piggy-backed into space by a mother craft before taking off, doing its business at up to 8 times the speed of sound (9500 kilometres per hour) and landing on its 'tricycle wheels' like a normal plane on a runway. It would then repeat this up to 25 times.

But in 2001, following budget cuts, changing space policy and NASA, the military and private engineers not being able to get the planes to work, the *X-34* program was scrapped.

According to Tyler Rogoway and Joseph Tervithick of website The Drive, despite a brief flurry of excitement in 2013 when it looked like the *X-34*s might be making a comeback, they now sit abandoned in a residential backyard in Florida. An attempt to move them to an aeronautical museum apparently got bogged down in paperwork so rather than orbiting the Earth at supersonic speeds, the *X-34*s collect bird poop and rust away.

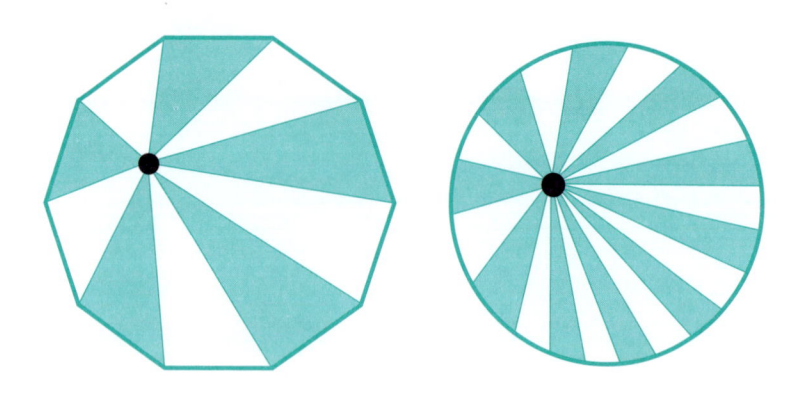

Taunting Tanton

James Tanton is a fiendishly talented mathematician, maths teacher and author.

Born in Adelaide in 1966, he got into his stride in America where he's taught maths in both high schools and universities for the past 30-odd years. His books and videos have delighted (yes, that's *exactly* the right word!) hundreds of thousands of people (if not millions).

Here's a nifty little Tanton twister. First thing you need to do is choose a point inside a regular even-sided* polygon and form evenly-spaced triangles (like the diagram). Then, colour them white and green (in alternating patterns). What can we see when it comes to the sum of green areas and the sum of white areas? Is there a comparable result for a point inside the circle?

I can assure you this puzzle has little to do with beach umbrellas or pizza!

*To say a polygon is even-sided means it has 4 or 6 or 8 or 10 or any even number of sides. To say the polygon is regular, means all of its sides are of equal length.

So in considering the circle, take the points around the side to be spaced equal distances apart. Again, assume there are an even number of points.

El Cap in hand

On 14 January 2015, Tommy Caldwell and Kevin Jorgeson reached the top of the famous El Capitan peak in the Yosemite National Park in California.

The spectacular vertical rock formation is about 900 metres from base to summit and made almost entirely of granite. It's a popular climbing spot, but Tommy and Kev did it as a 'free-climb' which meant while they used ropes and devices to stop them from falling, they did all the actual climbing themselves.

The climb was considered by many impossible and was made no easier by the fact that Tommy is missing a finger on one hand.

If you're interested in seeing more about their climb, check out the film *The Dawn Wall* which captures their amazing, and perhaps insane, 19-day athletic feat.

Some of you may remember Yosemite Sam, the cartoon character from the Warner Bros' *Looney Tunes* and *Merrie Melodies* series. No? Head to YouTube. Along with Elmer Fudd, Yosemite Sam generally has it in for ol' Bugs Bunny.

What the pluck?

The theorbo was an Italian lute that was banging popular in the Baroque period from 1600 until 1750.

If you're wondering what a 'lute' is, well, it's any one of a family of European stringed instruments (some with frets, others without) with a deep round back enclosing a hollow cavity. Kinda like a small guitar.

Anyway, the most common theorbo had 14 strings in vertical 'courses' running down the neck of the instrument, though some of the courses could be 'double-strung'. Unlike a lute, there was a second neck for the deeper sounding bass strings.

While the most common number of courses was 14, there were 15-course theorbos and, if you really loved those courses, the occasional 19-string theorbo was also known to get a pluck now and then.

The oldest string instruments, called lyres, were used in Mesopotamia around 2500 to 3000 BCE.

Thorny issue

The term 'spinster' is reasonably well known.

It refers to a woman who has never been married. Now you may well argue that it's no one's business whether a woman has been married or not and a special word to describe it is a little bit on the 'old-fashioned and slightly offensive' side. I for one would not stop you thinking that.

But it turns out not all never-married women are spinsters. From the mid-1600s in the newly-formed colonies of America the term spinster only applied to women aged 23 to 26. Once a never-married woman passed the age of 26, she graduated to ... wait for it ... 'thornback' status.

When 26-year-old Sophia Benoit, from Los Angeles, California, found this out and announced it on Twitter it's fair to say the internet melted. A 'thornback' is a sea animal with thorns down its spine, so I'm assuming the term was not a compliment.

Some of the best Twitter comments included:

'I have been looking for a great name of a girl's band since college ... JUST found it 40 years later ... The Thornbacks ... freakin marvellous!'

'Thornback sounds like a type of predatory dinosaur and that's badass.'

and *'I was a thornback for a full 10 months! I feel like I missed out on using my superpower.'*

Ah, the beauty of free speech.

We humans are a weird mob.

If you're unmarried at age 25 in Denmark, you can expect to get cinnamon thrown on you for your birthday ...

ADAM SPENCER

Crime peaks in the city of Manchester at a temperature of 18° Celsius

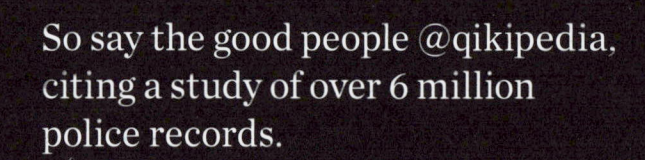

So say the good people @qikipedia, citing a study of over 6 million police records.

This is roughly the average high temperature in Sydney in August.

... Just sayin' ...

Judgey McJudgeface

Whether we like it or not, most of us judge others.

Hey, don't judge me — it's human nature.

But when we take it a step further and view other people as somehow 'less human' we're in real trouble. At the moment in many countries around the world (America, here's looking at you), a troubling dehumanising trend has been brewing. By this I mean the tendency for one person to view another person (usually of a different ethnicity) as 'less than human'. Social scientists have long linked such tendencies as key in the escalation of tension, conflict ... and worse.

So why do we do it and how can we evolve not to?

Nour Kteily, a psychologist from Northwestern University in Chicago, has come up with a tool that gauges people's desire to dehumanise by ... simply asking them the question directly.

Surveying a random group of Americans, Kteily found that Swiss, Japanese, French, Australian, Austrian, and Icelandic people were seen to have largely indistinguishable features from themselves, while Arabs, Mexican immigrants and Muslims were ranked as 'less evolved'.

It's suggested we view others as 'less human' because of the misplaced belief they want to do us harm. To counter this, we need to 'short circuit' the belief. Even more effective — but harder to accomplish — is getting to know people who are different from us; what's known in psych circles as the so-called 'contact hypothesis'.

I don't know about you, but I'm going to randomly smile a whole lot more when I'm out and about and see what happens! Surely it can't hurt.

What are the odds?

Warning: this story contains some pretty grisly numbers ...

In 2016, violence claimed the lives of some 560,000 people around the world. About 385,000 people were victims of intentional homicides, some 99,000 were casualties of war, while the remainder died in unintentional homicides and 'legal' executions. Overall, that corresponds to about 7.5 violent deaths per 100,000 population.

So how do you avoid becoming a horrific statistic?

Well, we can't predict the future, but taking precautions before visiting some of the most dangerous parts of the world is a good start. Tragically, in 2016, Syria, El Salvador, Venezuela, Honduras and Afghanistan had the highest violent death rates. Nigeria, Yemen and Somalia were close behind. Promise me you'll be careful if visiting any of these places.

Then of course you could spend more time in the *safest* places on the planet. Liechtenstein regularly records no murders in a year (but its population is pretty small). Iceland, New Zealand, Austria, Denmark and Singapore similarly all score well when it comes to safety.

And Australia? Glad you asked.

We generally make the top 20 safest countries. Behind Finland, Portugal, Canada and Spain ... but ahead of Germany, France, the UK and the US.

'An equation for me has no meaning unless it represents a thought of God.'

— Srinivasa Ramanujan

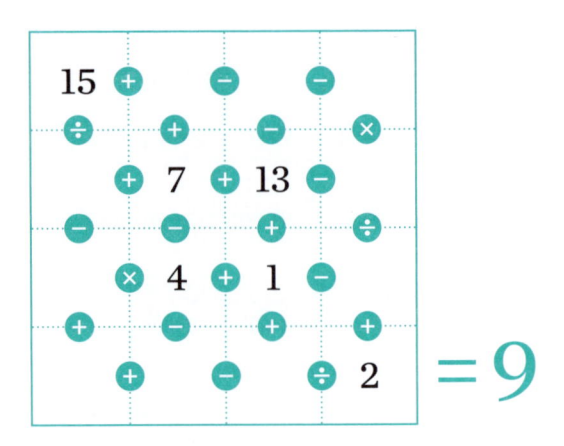

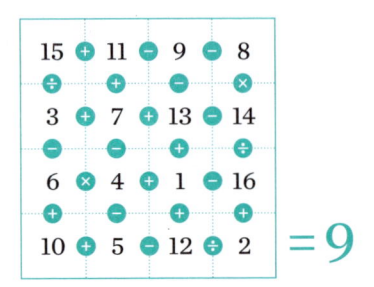

Operation!

As always in *Numberland* (as in the 'real world') order of operations matters ...

Which means it's worth bearing in mind that Pandas Eat Milk Duds and Skittles.* Just like you'll eat these puzzles for breakfast! Okay, well let's not get too hasty. They're no walk in the park.

For each 4 × 4 grid, enter the numbers 1 to 16, once and only once, so that each row and column equals the target number.

There's an example above to kick things off, and these first few puzzles have plenty of hints to ease you into the process. But later in the book, the number of hints will reduce and you'll just have to sharpen your pencil, strap your thinking cap on and knuckle down!

*See page 270 if this doesn't make sense to you.

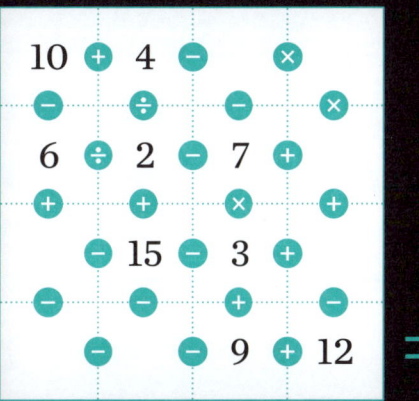

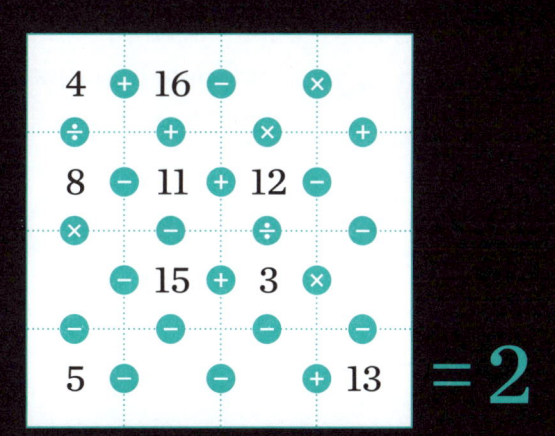

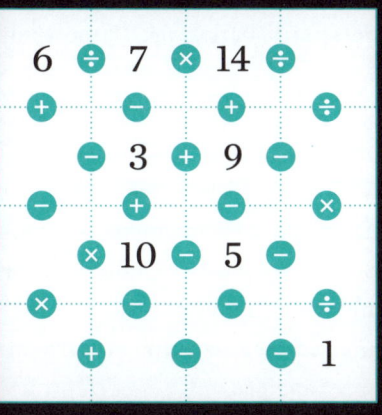

NUMBER LAND

Puzzle 1

```
10  +   4   -       ×
 -      ÷       -   ×
 6  ÷   2   -   7   +
 +      +       ×   +
        -  15   -   3   +
 -      -       +   -
        -       9   +  12
```
= 1

Puzzle 2

```
 4  +  16   -           ×
 ÷      +       ×       +
 8  -  11   +  12   -
 ×      -       ÷       -
        -  15   +   3   ×
 -      -       -       -
 5  -       -       +  13
```
= 2

Puzzle 3

```
 6  ÷   7   ×  14   ÷
 +      -       +       ÷
        -   3   +   9   -
 -      +       -       ×
        ×  10   -   5   -
 ×      -       -       ÷
        +       -       1
```
= 3

YOU ARE HERE
57

Gimme a break!

Sport is tough. It's punishing. It takes it out of the players. And the officials. And that's just lawn bowls.

But not all sports reward players with equal 'time out'. No, my dear jocks and jockettes, while we give thanks for the quarter time (6 minute) and halftime (20 minute) breaks in AFL — refreshments, anyone? — if you're on a field playing American football you only get 2 minutes and 12 minutes (unless you're playing in a Superbowl and Taylor Swift is on fire). Better in basketball, where you get 15 minutes at halftime. Particularly cute is that in most basketball worldwide, the quarter and three quarter-time breaks are 2 minutes. But in the American NBA they are 130 seconds. More time to sell shoes?

Nevertheless, it's worse in rugby league where you get a single break of 10 minutes.

Other sports of course don't even bother with that 'halftime' nonsense: professional ice hockey games consist of 3 × 20-minute periods, broken up with 17-minute intermissions which allow, among other things, time to repair the playing surface. 17 minutes! Hey, have you ever tried to wash blood off ice?

AT, okay!

The Appalachian Trail (or AT) is a marked hiking trail which doubles as a gruelling marathon course. It's also a testament to the extraordinary machine that is the human body.

In July 2018, a 28-year-old dentist from Belgium by the name of Karel Sabbe completed the 3522 kilometre marathon through Georgia and Maine in the US in just 41 days, 7 hours and 39 minutes. This beat the previous best-ever Joe 'Stringbean' McConaughy's record by an incredible 4 days.

Sabbe averaged around 85 kilometres per day — the same as two regular marathons — for more than 41 days straight. Most hikers take 169 days, notching up 22 kilometres per day, to complete the trail.

Naturally, Sabbe was stoked. 'Nobody had averaged more than 50 miles [80 kilometres]. More than proud, I feel privileged for having lived these incredible adventures. It was a blast from start to finish!' A *blast*, you say, Karel?

Unlike McConaughy, Sabbe didn't do it alone, and his friends, wife, brother-in-law and parents supported him, making sure he ate his 10,000 calories a day.* And brushed his teeth.

* It depends on your age, sex and body type, but roughly speaking, adults should only eat 2000 calories per day (or 8700 kilojoules).

'At first, you crave carbs, and then that shifts to fat, and I was eating pizza, potato chips, lots of candy, M&Ms, granola bars, and things like that,' Sabbe said. 'I usually ate on the uphills so I didn't waste time.'

He was also helped by fellow ultramarathon runner Scott Jurek who'd completed the 'AT' himself in 45 days, 22 hours and 38 minutes. Jurek managed an amazing 136 kilometres on his last day ... and had to get up at 3.30 am to get it done!

Besides the Appalachian Trail record, Karel Sabbe also holds the record for the Pacific Crest Trail, which he completed in 52 days, 8 hours and 25 minutes in 2016. He is the only person to hold both records at the same time.

Karel, you put the legs into legend.

NUMBERLAND

iFreeze ...

Does the thought of being locked out of your smartphone (or smart watch or smart device) for even a few hours frighten you?

Then spare a thought for poor Evan Osnos, whose 3-year-old kid locked him out ... for 48 years.

Yep, after repeatedly trying to unlock the passcode protected device, Evan received the message no modern parent ever wants to receive: *Your device is disabled ... try again in 25,536,442 minutes.*

Say what? On a lot of devices, if the incorrect passcode is entered more than 6 times (and you inadvertently repeat this process many more times) you'll get that sort of horrifying message. But don't worry, it won't truly lock you out for that amount of time. It's just toying with you. Actually, it's telling you (or your 3-year-old kid) to stop toying with it by entering passcodes and take a breather.

In case you were wondering, it's estimated there are around 18 million smartphone users currently in Australia. That's a pretty impressive statistic considering there are only just over 25 million of us in total. We're ahead of the pack too: we got to 90% of the population with a smartphone in 2018 — the rest of the world won't catch up till 2023.

YOU ARE HERE

61

A bee in maths

While we humans are pretty clever when it comes to counting, don't count out our non-human friends in the natural world. Adrian Dyer from RMIT has been studying bees and has made some startling discoveries.

While the bee brain is considerably smaller than the human brain (1 million or so neurons compared to our 86 billion) and its architecture massively different, Dyer and his team have shown that bees may very well be able to count.

How? Well, Dyer politely asked 14 hungry bees to enter a Y-shaped maze. Inside the maze, they would see from 1 to 5 shapes that were either blue or yellow. The bees had a choice to fly to the left or the right side of the maze, with one side containing one more element and the other containing one less.

If the shapes were blue, the bees needed to add an element; if the shapes were yellow, they had to subtract. If they got it right, they were rewarded with sugar water. Get it wrong, they got a nasty tasting quinine solution.

After several hours of training, Dyer and his team repeated the challenge, this time without the punishment or reward. What happened? Well, you guessed it, the bees chose the correct answer 60 to 75% of the time!

As to how and why the bees managed it, Dyer suggested

they've either evolved this ability because they have to process a lot of complex information going from flower to flower, or their brains are far more 'plastic' than we first thought.

To put it in perspective, these bees were doing a similar level of maths as an average child of 4 or 5.

But not all animals are competing with us when it comes to doing maths. Back in the early 1900s, following Charles Darwin's extraordinary work, people were fascinated by animal intelligence. Along comes 'Clever Hans', an Orlov Trotter horse, who was allegedly taught to add, subtract, multiply and divide ... and a whole host of other incredible tasks. You can read more about this clever clogs (or hooves) on page 298.

Turns out it wasn't necessarily so, bro. In 1907, a psychologist named Oskar Pfungst proved that Hans wasn't actually performing these amazing feats ... he was just responding to cues from his trainer. Which is still pretty impressive. But maybe not as impressive as those RMIT bees.

ADAM SPENCER

Would you believe it's legal to send live scorpions in the mail in the US?

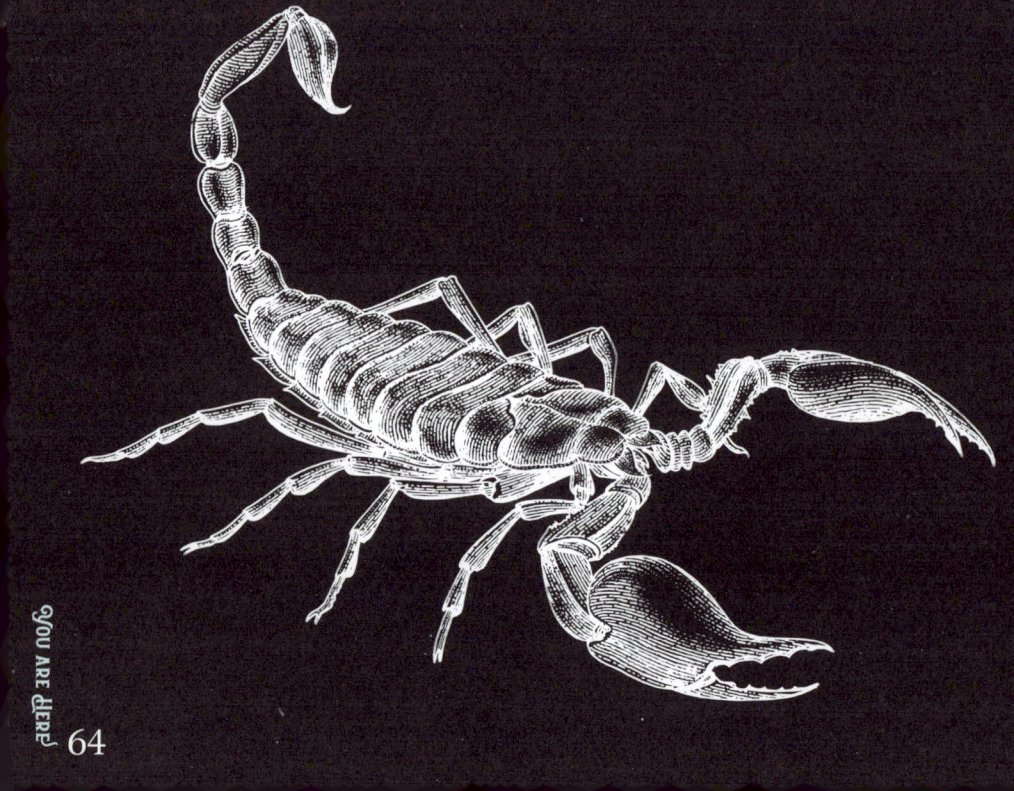

Well, it is, just as long as the scorpions are in a box labelled 'live scorpions' which is in a second box ... also labelled 'live scorpions'.

But just because it's legal, I don't recommend you do it!

Barking Barkley

Finished with the Comrades marathon?

Then why not turn your attention to the punishing Barkley event held each year in the aptly named Frozen Head State Park near Wartburg in Tennessee?

Sure, it's called an 'ultra' marathon, but I think 'hyper' is a more appropriate name.

Each year, this punishing '100-odd mile' (160 kilometre) monster chews up participants and spits them right out. I'm serious. Dubbed the marathon that 'eats its young', for the second year running in 2019, not one single person managed to finish.

That's right, the Barkley Marathons ate *all* of the competitors. The 5 loops of the monstrously mountainous course must be completed in less than 60 hours. That is brutal. Compare this to Sydney's City2Surf, which is a mere 8.7 miles (14 kilometres) and usually run and won in around 41 minutes (though the record is held by Steve Moneghetti who smashed out a 40 minutes and 3 seconds time back in 1991).

In case you were wondering, the Barkley event was first staged in 1986, and since then only 15 people have *ever* finished it.

Rollover, GPS!

The good ol' GPS (global positioning system) really is a triumph of scientific brilliance.

Around 30 satellites form what's known as the 'GPS constellation'. They orbit the planet roughly 20,000 kilometres overhead with each satellite completing two orbits of the Earth every day.

Down on the ground, GPS receivers combine the microwave signals sent from the satellites in a process called 'trilateration' to determine the position of an object (or you with your GPS-equipped device).

So far so perfect, right?

In order to calculate precise locations, the satellites include a time stamp with their microwave signals. This time stamp includes a 'week number' which is sent as a string of 10 binary digits — 1s and 0s — and as we all know, there are only 1024 (that is, 2^{10}) possible combinations of said 10 binary digits. This means that one GPS 'term' equals 1024 weeks (19.7 years). And what happens at the end of that term? Well, you guessed it again: the GPS system 'rolls over'. Just like Y2K.*

The first GPS systems began on 6 January 1980, so the first rollover happened on 21 August 1999. But you're quite right, my friends, we didn't suddenly lose our way (and the Earth didn't end, either). Software engineers had time to prepare and so most systems merrily rolled over and welcomed the new millennium with nary a hangover to be seen. Crisis averted.

* The Y2K bug was a computer flaw, or bug, that may have caused problems when dealing with dates beyond 31 December 1999. The flaw, faced by computer programmers and users all over the world on 1 January 2000, is also known as the 'millennium bug'. The letter K, which stands for kilo (a unit of 1000), is commonly used to represent the number 1000. Y2K stands for Year 2000. Sceptics believe it was barely a problem at all. Others say it stands as a great example of focused effort and problem-solving worldwide. Either way, we were certainly worried at the time!

The *flying kangaroo*

We've come a long way when it comes to international airline travel.

According to the Qantas website, an 84-year-old outback pioneer named Alexander Kennedy was the first Qantas passenger on a scheduled flight. He had agreed to subscribe some cash and join the provisional board provided he got ticket No.1. His flight, on 2 November 1922, was on the Longreach-Winton-McKinlay-Cloncurry section of the inaugural mail service from Charleville to Cloncurry. Meanwhile, the first jet service was on 29 July 1959 from Sydney to San Francisco via Nadi and Honolulu.

Just last year, Qantas broke the record for a Perth to London flight when its 787 Dreamliner made the journey in 16 hours and 29 minutes (almost an hour quicker than the scheduled 17 hours and 20 minutes) at an average speed of 938 kilometres per hour.

And in case you were wondering, over the next 12 months, the most popular movie to watch on board that flight was ... *Mission Impossible*! Of course it was. But wait, there's more. On the first anniversary of the flight, Qantas revealed one particular passenger had stayed in his business class suite the whole trip — without a single toilet break. Hmm. Not the recommended way to travel but hey, we're talking records here, not medical best practice.

And if you're still wondering (c'mon, work with me here) seat 56F holds the record for the most in-flight entertainment watched (in a year) with 9134 hours.

I think there's a TV show in this.

Shine bright like a diamond

The Graff Lesedi La Rona diamond clocks the scales at a whopping 302 carats (around 60 grams).

It came from a 1109-carat rough diamond found in Botswana in 2015. The English jeweller Laurence Graff paid US$53 million for it and spent 18 months cutting, polishing and probably just gazing in awe at the thing.

It's a big rock, no doubt about it, but it pales in comparison to the largest diamond ever unearthed and the family of baby super-Ds* it created.

The Cullinan diamond was found in South Africa in 1905. It weighed 3106 carats (621 grams) and was named after the mine manager at the time, Thomas Cullinan. It was put up for sale in London but no one bought it so in 1907 the Transvaal Colony government gave it to King Edward VII who sent it off for cutting and polishing.

This massive rock gave birth to various smaller diamonds, the largest of which is called Cullinan I or, more commonly, the Great Star of Africa. At 530 carats (106 grams), it's the biggest clear-cut diamond in the world, and features in the head of the Sovereign's Sceptre with Cross.

The next biggest diamond, you guessed it, the Cullinan II (or the Second Star of Africa), comes in at a very respectable 317 carats (63 grams). It's mounted in the Imperial State Crown. Both of these Cullinan diamonds are part of the English Crown Jewels. Queen Elizabeth privately owns several other massive rocks ... not to mention a few other treasures.

It might just be me, but a lot of kings and queens seem to do alright in the receiving gifts department.

* Not an official term — I just made 'baby super-ds' up.

'Listen to the great Beatles records, the earliest ones where the lyrics are incredibly simple. Why are they still beautiful?

'Well, they're beautifully sung, beautifully played, and the mathematics in them is elegant.'

—Bruce Springsteen

The most expensive hyphen in history

Look, I've got to be honest. I know the pain of finding a typo in a boko. But I'm pleased to say that none of mine ever led to a taxpayer bill quite this high.

When NASA's *Mariner I* rocket was launched on 22 July 1962 to do a pass of Venus to collect scientific data — NASA's first planetary mission, I should note, during a particularly tense period in the US vs Russia 'space race' — there was a lot riding on it, figuratively, at the very least.

Though fortunately unmanned, the cost of the rocket came to approximately US$80 million which, in today's dollars, equates to about US$673 million. So it was with some mixed feelings, I'd imagine, that a NASA range-safety officer flicked a switch and blew it up 293 seconds into the launch after it veered off its correct trajectory.

Though there were conflicting reports, many claimed that a single, misplaced hyphen in the rocket's code was to blame. Famed sci-fi writer Arthur C. Clarke called it 'the most expensive hyphen in history'.

Fortunately NASA had *Mariner II*, a back-up rocket, which went on to complete its mission safely.

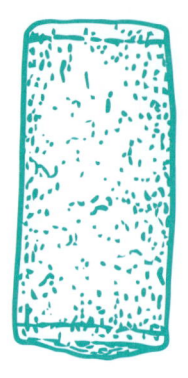

A corker of a riddle

Imagine I hand you a wine bottle cork, a glass and a match.

Now, I also give you a plate which is covered in just enough water to cover it entirely.

Gee, thanks, Adam!

What's the point of this eclectic collection of *stuff*, you ask? Well, another riddle, of course!

Tell me, is there a way of filling the glass with water without moving or even touching the plate, using only the objects listed?

Have a think, then, if you need to, grab the items listed and have a play.

But remember, kids, matches are dangerous. Ask Mum or Dad to give you a hand before trying it.

Same goes for you, pyro mums and dads: matches are dangerous. Be careful.

Answer's at the back o' the book, as always.

Happy birthday!

From Dimitri Mendeleev's OG* periodic table of 1869 (reproduced here in *Popular Science Monthly* Volume 59) ...

			Ti....50	Zr...90	?....180
			V.....51	Nb..94	Ta...182
			Cr....52	Mo..96	W....186
			Mn....55	Rh..104.4	Pt...197.4
			Fe....56	Ru..104.4	Ir....198
			Ni,Co..59	Pd..106.6	Os...199
H.....1			Cu.....63.4	Ag...108	Hg..200
	Be....9.4	Mg...24	Zn....65.2	Cd...112	
	B.....11	Al....27.4	?......68	Ur...116	Au...197
	C.....12	Si....28	?......70	Sn...118	
	N.....14	P.....31	As....75	Sb...122	Bi....210
	O.....16	S.....32	Se....79.4	Te...128?	
	F.....19	Cl....35.5	Br....80	I.....127	
	Na....23	K.....39	Rb....85.4	Cs...133	Tl....204
		Ca....40	Sr....87.6	Ba...137	Pb...207
		?.....45	Ce....92		
		?Er...56	La....94		
		?Y..60	Di....95		
		?In...75.6	Th....118		

... To the current iteration featuring Og itself ...

You've come along way, baby! Still looking fresh in 2019 at 150 years of age. Happy birthday, periodic table!

*That's 'Original Gansta' for the uninitiated ...

Long losing streak

Some reports reckon the Washington Generals basketball team beat the famous Harlem Globetrotters 6 times, but the Generals' website says there were just 3 victories, one in 1954, one in 1958 and one in 1971.

Whatever the case, the 1971 game is the one everyone remembers. Why?

Well, playing as the 'New Jersey Reds', the Generals had had a pretty rotten run — they hadn't won a single game in, wait for it, 2495 attempts!* But on 5 January 1971 in Martin, Tennessee, all that changed.

As the awesomely entertaining Globetrotters went about wowing the crowd with their tricks, they lost track of the game — and the score. Suddenly, they found themselves behind by 12 points with just 2 minutes on the clock.

As the seconds counted down, the Generals' 50-year-old player/coach Louis 'Red' Klotz hit the winning basket. The Globetrotters' Meadowlark Lemon tried but could not respond and the final buzzer announced the unthinkable — the Generals had beaten the Globetrotters 100–99!

The fans were not happy. 'They looked at us like we killed Santa Claus,' Klotz said. Kids were crying, mums and dads were trembling. But the victory belonged to the Generals — and it was made all the sweeter as they haven't repeated it since.

*In all fairness, that's sort of what they were meant to do since they were created purely to act as a stooge in Globetrotter games ...

Squash player Jahangir Khan holds the record for the longest winning streak in *any* professional sport …

The Pakistani supremo won an astonishing 555 matches over a period of 5 years between 1981 and 1986.

ADAM SPENCER

Sixty per cent of people can't make it through a 10-minute conversation without lying ...

And that's the truth, Ruth, according to research from University of Massachusetts psychologist Robert S. Feldman, published in the *Journal of Basic and Applied Social Psychology*.

Subjects told an average of 2 to 3 lies, no matter their sex; however the study also found that lies told by men and women differ in their content. Feldman noted that 'women were more likely to lie to make the person they were talking to feel good, while men lied most often to make themselves look better'.

Phfttt. Yeah, right.

Hey, while I think of it, have I ever told you about that time I kicked the winning goal in a Grand Final for the Swans?

Want a million bucks?

In 3 space dimensions and time, given an initial velocity field, there exists a vector velocity and a scalar pressure field, which are both smooth and globally defined, that solve the Navier–Stokes equations.

If you can prove — or disprove — that statement, you can claim a million-dollar Millennium Prize from the Clay Mathematics Institute of Cambridge, Massachusetts (CMI). And they're US bucks, so given the exchange rate, it's worth sharpening your pencil.

The bad news is that this problem is considered one of the world's least understood at a theoretical level. The good news — payola aside — is that if you can crack it you'll contribute to solving all manner of humanity-helping questions regarding turbulence, which in turn might help with everything from space travel to climate modelling.

It's all to do with describing the motion of a Newtonian fluid* in space, taking into account a range of variables such as kinetic energy.

So there's your hint. Get started.

---

* Newtonian fluids include water, air, alcohol and thin motor oil — they all adhere to the simplest mathematical model of a fluid accounting for viscosity.

Non-Newtonian fluids include non-drip paint, which becomes thinner when sheared, custard and the awesomely named *ooblek* which actually becomes stiffer when it's 'vigorously sheared' — or punched.

Get thee to YouTube for some nifty videos of it on subwoofers and in other interesting scenarios. Or head to the fridge, dig out that jar of old, cold caramel ice-cream sauce and give it a very quick poke with your finger — you'll find it now behaves more like a solid than a liquid.

How about 6 million bucks?

Well, only 5 if you just nutted out that pesky Navier-Stokes problem.

But the balance is up for grabs, my friend.

To celebrate 'mathematics in the new millennium' the Clay Mathematics Institute established 7 Prize Problems.

They were announced on 24 May 2000 at the Collège de France in France having been conceived 'to record some of the most difficult problems with which mathematicians were grappling at the turn of the second millennium; to elevate in the consciousness of the general public the fact that in mathematics, the frontier is still open and abounds in important unsolved problems; to emphasise the importance of working towards a solution of the deepest, most difficult problems; and to recognise achievement in mathematics of historical magnitude'.

Hey, I didn't say that winning this sort of coin would be easy. Or very likely. But just by *knowing* about these problems you'll be metaphorically richer!

Consider for a moment the depth and breadth of human achievement over the past few thousand years. From the extraordinary — putting a man on the Moon surely qualifies — to the seeming mundanity of modern plumbing, humankind has traversed such an enormous field of learning during its brief time here on Earth. But we still have so much to learn. As the CMI says 'the frontier is still open and abounds in important unsolved problems' which, once solved, will open our eyes to further horizons still … and almost certainly spur even more unsolved connundrums for us to ponder.

Wondering what the 7 problems are? Grab your thinking cap and flip the page …

The 7 Millennium Prize Problems

Here we go! The 7 Prize Problems.

You might recall we've looked at a few of them in some of my other books (available where all good books are sold ...) and we'll look at more in these pages.

When formally expressed, they are jam-packed with notation and symbols that can make even those with the strongest of mathematical stomachs need a little bit of a sit down. So, instead I present to you, the 'plain English' versions of these million dollar babies.

Yang–Mills and Mass Gap

Experiment and computer simulations suggest the existence of a 'mass gap' in the solution to the quantum versions of the Yang-Mills equations. But no proof of this property is known.

Riemann Hypothesis

The prime number theorem determines the average distribution of the primes. The Riemann Hypothesis tells us about the deviation from the average. Formulated in Riemann's 1859 paper, it asserts that all the 'non-obvious' zeroes of the zeta function are complex numbers with real part $1/2$.

P vs NP Problem

If it is easy to check that a solution to a problem is correct, is it also easy to solve the problem? This is the essence of the P vs NP question. A typical NP problem is that of the Hamiltonian Path Problem: given any map of cities and roads between them, is it possible to visit every city without visiting a city twice?

As I said, the 'official 'statements of the problems involve some truly intense language and notation that you can look up online at your leisure.

Navier–Stokes Equation

This is the equation which governs the flow of fluids such as water and air we met on the previous page. However, there is no proof for the most basic questions one can ask: do solutions exist, and are they unique? Why ask for a proof? Because a proof gives not only certitude, but also understanding.

Hodge Conjecture

The answer to this conjecture determines how much of the topology of the solution set of a system of algebraic equations can be defined in terms of further algebraic equations. The Hodge Conjecture is known in certain special cases, for example, when the solution set has dimension less than 4. But in dimension 4 it is unknown.

Poincaré Conjecture

In 1904, the French mathematician Henri Poincaré asked if the 3-dimensional sphere is characterised as the unique simply-connected 3-manifold. This question, the Poincaré Conjecture, was a special case of Thurston's geometrisation conjecture. Perelman's proof tells us that every 3-manifold is built from a set of standard pieces, each with one of 8 well understood geometries.

Birch and Swinnerton-Dyer Conjecture

Supported by much experimental evidence, this conjecture relates the number of points on an elliptic curve modulo a prime to the rank of the group of rational points. Elliptic curves, defined by cubic equations in two variables, are fundamental mathematical objects that arise in many areas: Wiles' proof of the Fermat Conjecture, factorisation of numbers into primes, and cryptography, to name 3.

The Poincaré Conjecture was solved by the brilliant but enigmatic Russian maths whiz Grigori Perelman in 2006.

But Perelman refused to accept the $1,000,000 or the highest accolade in mathematics, the Field's Medal. 'I don't want to be on display like an animal in a zoo,' he said.

The money was directed to a scholarship for young genius mathematicians instead.

Stayin' alive

Did you know that the song 'Stayin' Alive' by the Bee Gees can save someone's life?

Literally.

It turns out that this disco-era earworm averages 103 beats per minute, which is almost exactly the rhythm you need to keep if you're giving someone emergency CPR.

But don't just take my word for it. The song was chosen by the American Heart Association because of its widespread popularity and even included in its official advice to 'call 911 and push hard and fast in the centre of the chest to the beat of the classic disco song "Stayin' Alive".'

There have since been a number of reports about the song doing just like the title says — keepin' people alive.

However, BBC News reports that a number of studies suggest that while this trick does indeed help ensure people maintain the correct rate, the pressure they apply when using the song is frequently too low.

So, by all means sing along to the Bee Gees' classic, but perhaps bear in mind another 90s megahit as you go about administering potentially life-saving CPR. You remember Salt-N-Pepa ...

Come on, Spinderella, cut it up one (more) time ... 'push it *real* good'!

The beautiful game

Not for nothing is football — or as *some* call it, *soccer* — known as the beautiful game.

I could carry on about the artistry of the footwork on field, or point you to the shape of the ball itself (a truncated icosahedron, since you asked) but for greater proof of the beauty in the game you'd be hard pressed to beat this.

The first Ivorian Civil War in Africa began in 2002 and ended in 2007 ... with a soccer match when the local Ivory Coast team qualified for the World Cup.

Live on television, the team dropped to their knees and beseeched the nation to put aside their differences. They then arranged for a qualifier for the African Cup to be held in a rebel-held city. One thing led to another and the two sides reached a peace agreement.

Sadly, it was short-lived and violence once again erupted in 2011 in the wake of a disputed election, killing over 3000 people. But soccer would once again play a part in ending the conflict, with Chelsea striker Didier Drogba playing an instrumental role off the field in bringing the opposing factions back to the negotiating table.

'When Ivory Coast is playing, the country is united,' he said. 'People who don't talk to each other, when there is a goal they celebrate together ... we are playing football, we are running behind a ball, and we managed to bring people together.' Legend.

But soccer hasn't got a perfect record ...

Soccer had a smaller victory in September 1967 when the two opposing sides in the Biafran War declared a two-day truce ... to watch Pelé and his touring Santos team play two exhibition matches.

The (not so) beautiful game

Soccer has, in fact, almost directly caused another war.

The year was 1969 and the 'Football War' was fought both on and off the field between Central American countries El Salvador and Honduras.

A number of simmering tensions were brought onto the pitch during the best-of-3 grudge match to decide who would qualify for the World Cup.

The first game in Tegucigalpa went 1-0 to Honduras with some relatively minor disturbances; however things took a turn for the worse during the next game in San Salvador. The visiting Honduran players were greeted with rotten eggs, dead rats and worse at their hotel. Their flag and anthem were mocked and their fans beaten in the stands. The Honduran coach admitted '[we] had our minds on getting out alive. We're awfully lucky that we lost'. 3-0 El Salvador.

Tensions worsened still before the deciding third game, held in Mexico, with the press stoking the frenzy. On the day of the match, 27 June, Honduras broke off diplomatic relations with its neighbour. El Salvador went on to eventually triumph 3-2 in extra time. And a week later, they invaded Honduras, sparking a conflict that would last only until 20 July, but claim up to 2000 lives and leave more than 100,000 refugees.

Although troops withdrew from El Salvador that August, it would take some 11 years before a peace treaty between the two nations was reached.

Save 0.2 seconds with every 5 kilograms lost on the F1 diet!

It's not just jockeys, boxers and models who have to starve themselves before a gig.

Spare a thought for all the (model-dating, fast-car-driving, bajillion-dollar-earning) Formula One drivers out there who are precisely monitored and absolutely not allowed to carry any extra weight.

Not only are the cockpits of the cars extremely tight fits, it's estimated that an additional 5 kilograms of weight can add a whole 0.2 seconds per lap. That's an extraordinary amount in a sport where races are won by fractions of a second.

So if you want to throw your helmet in the ring as an F1 champ, prepare to undergo some pretty tough conditions. You'll need a target weight of 60—65 kilograms and, if you want to follow in the footsteps of, say, former British F1 driver Jenson Button[*], be prepared to starve, sweat and triathlon your way to raceday weight.

Oh, and lay off the pizza ... big time.

[*] Jenson Button won the F1 World Championship for the first and only time in 2009 with the then Brawn GP team. During his career, he started 306 races, winning 15 and getting onto the podium 50 times.

Did someone say 'free pizza'?!

Thought that might get your attention.

Keen readers of my books will know the humble pizza is always a hot topic — not only because of the DQ — that's deliciousness quotient (okay, I made that definition up) — but also because of how well it lends itself to mathematical problems and puzzles.

Say you're splitting pizza with friends and you want to know what size will give you the most mozzarella for your money. Or the most dough for your dosh. Or the most crust for your coin. You get the picture. Well, you'd probably think you would get more pizza if you order a couple of 12-inch bad boys as opposed to just one 18-incher, right?

You'd be wrong! Say what? Yep, as I was recently reminded by the good folk at Fermat's Library, you will actually get more by splitting one 18-inch pizza instead of ordering two 12-inch pizzas.

How can it be? Well, an 18-inch pizza has 254 square inches of 'pizza', while two 12-inch pizzas only have about 226 square inches.

Using that old chestnut from high school geometry, that the area of a circle with radius r is πr^2 and noting that the radius is half of the diameter, we see that:

One 18-inch pizza's area = $\pi \, (18/2)^2 \approx 254$ inches^2

Two 12-inch pizza's area = $2 \times \pi \, (12/2)^2 \approx 226$ inches^2

Next time you're on the beautiful Central Coast of NSW, pop into the Lady Copa pizzeria* and put this theory to the test. Ask for my mate Franco — he'll look after you.

*With any luck, I'll be the guy in the corner eating my free lifetime's supply of Franco's pizza for helping a brother out.

Ker-Planck

What's the tiniest unit of time you can think of?

A second? A millisecond? Ha! We go *way* smaller, my friend.

'Planck time' (also known as a Planck second) was first suggested by the German physicist Max Planck way back in 1899. Herr Planck thought there were fundamental natural units for length, mass, time and energy.

He came up with these using only what he believed were the most fundamental universal constants: the speed of light, the Newton gravitational constant and his very own Planck constant.

A Planck time unit is the time that is needed for light to travel a distance of 1 Planck length in a vacuum, which is a time interval of approximately 5.39×10^{-44} s.

We've talked about powers of 10 before* to explain massively large numbers. Well, they can also explain incredibly small ones. Here, the negative sign in the power means that rather than a 1 with 44 zeroes after it, we are talking about a number that begins with 0.000... and zeroes all the way down to the 44th decimal place before finally reaching 539.

Why is this significant?

Well, physicists deal with cosmological forces that are both extremely big and terrifically small. As such, they need to be able to measure things accurately.

And Herr Planck's unit is the measurement of choice when it comes to small.**

---

* If you need a refresher, where better to look than your nearest fine purveyor of books where you can grab a copy of (PLUG!) my *Big Book of Numbers*!

** Our good friend Planck popped onto the radar again in 2019 after scientists voted at the General Conference on Weights and Measures in Versailles to redefine what officially constitutes a kilogram. Having been defined for more than a century as the mass of a chunk of platinum-iridium (AKA Le Grand K), held in Saint-Cloud, France, the kilogram is *now* defined by setting the Planck constant (a universal physical constant that measures the energy carried by a photon) as $6.62607015 \times 10^{-34}$ m² kg/s.

The significance of this renaming is immense. The kilogram is no longer linked to a physical object (which is obviously susceptible to damage, decay or 'miscalibration'), but is now tied to a fundamental, physical constant of the universe.

Regardless, I'll certainly be looking to shift a couple of these newfangled kilos to look great for summer ...

87

									= 99

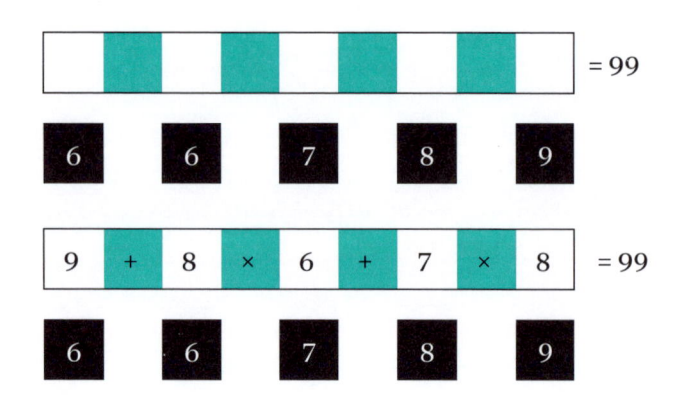

6	6	7	8	9

9	+	8	×	6	+	7	×	8	= 99

6	6	7	8	9

High scores

You've heard the phrase 'deceptively simple'?

Well, these puzzles might fit in that category. Simple to describe ... less so to solve.

The idea is to reach the goal number by creating an equation using the provided numbers and your own choice of operations. Just like the example shown above, really.

Numbers are placed in white squares. Operations (+, −, ×, ÷) are placed in green squares. Pandas Eat Milk Duds And Skittles* (or, rather, order of operations matters, for the uninitiated), and you can't use brackets!

Read the numbers off in order for your final score.

What is the equation that gives the highest score?

*See page 270.

Baby steps ...

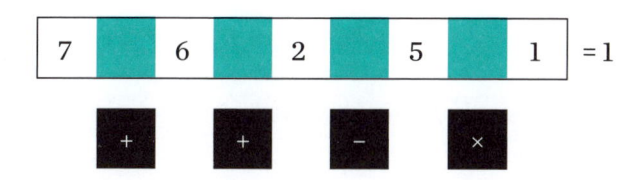

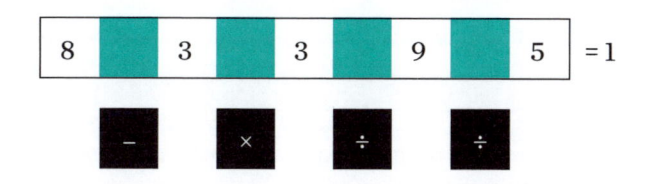

| 7 | | 6 | | 2 | | 5 | | 1 | = 1 |

+   +   −   ×

| 8 | | 6 | | 8 | | 5 | | 7 | = 1 |

−   −   ×   ×

To ease you into these beautiful (but challenging) problems, I'll start with some pretty substantial hints.

I've already given you the five numbers in the order that gives the high score in the equation, and I've also given you the four operators you'll need.

Your job is simply to work out the correct order for the four operators.

Trust me: it'll get a lot tougher by the end of the book ...

| 8 | | 3 | | 3 | | 9 | | 5 | = 1 |

−   ×   ÷   ÷

YOU ARE HERE

89

Jumpin' Japanium

True fans will remember our terrific trot through the elements in that bona fide 2015 bestseller *World of Numbers* (still available where all good books are sold).

And if you're reading this book in order (and hey, why wouldn't you?) then you'll likely recall the little birthday celebration we threw back on page 73!

Now, in spite of its achievements, our periodic table is lacking. In letters, I mean.

It could have been so different. If nihonium (a synthetic chemical element with the symbol Nh and atomic number 113) had been called japanium we would finally have the entire alphabet represented on the table — as it stands, J remains the only letter not featured. Instead, in 2017, they named the element after Nihon — the Japanese name for, you guessed it, Japan.

Well, that's not quite true. There is also no longer an element symbol that contains the letter Q. Previously, element 114 had the placeholder name 'ununquadium' and symbol Uuq, however it was officially renamed flerovium (Fl) in 2012 after its discovery in 1998. Tough break, Q.

Those same true fans will recall that the letter J was one of the last letters to come to the 'alphabetic table' too … what's the world got against J, J?

I've said it before: as a single diagram that summarises, through the beauty of numbers, a bulk of knowledge, the periodic table might just be humanity's most amazing achievement.

Home delivery

The scientific jury is still out, but new research from the University of Michigan suggests that home delivered meals might not be all that bad for the environment.

Say what? How about all that plastic, and cardboard, and other packaging; not to mention the costs of getting the goodies to your door?

Looking at the big picture — examining every stage of the food production process from farm gate to landfill — the researchers found that average greenhouse gas emissions were actually around 33% lower for home delivered meals than meals you buy at the supermarket.

The study acknowledged that home delivered meals often contain a lot of packaging, but worse still is the amount of food we buy at the supermarket ... but end up throwing out!

Yep, that extra packet of cheese/chicken thigh/jar of *ooblekky* caramel sauce that gets forgotten and passes its use-by date is more damaging to the environment because of all the energy and materials used in its original production and the energy and cost of then disposing of it.

Home delivery it seems tightens the process, resulting in less wastage — and lower emissions.

As I say, the research is still evolving ... so watch this space.

Oh, and *bon appétit*.

Onya Onsen

A unique and compelling aspect of travelling around Japan is sampling the various *onsen* (hot springs) and *sento* (community bathhouses) available.

Both terms refer to public baths, the difference being that an *onsen* is fed by natural geothermal springs while a *sento* (generally) uses heated tap water.

To maintain the distinction, the *Onsen Law*, established in 1948, prescribes that an *onsen* must contain either water over 25° Celsius or at least one of 19 specific natural chemical elements, like sodium chloride, calcium chloride, magnesium, iron and sulphur.

Another restriction, unofficial yet still regularly enforced, is against customers with tattoos.

On a 2019 visit to a wonderful *onsen* in Tokyo, I found that a couple of discreetly placed bandages did the trick.

Thankfully I wasn't covering a full neck and sleeve job!

Pi-ku

While we are in the land of the Rising Sun, let's talk poetry.*

To understand this next piece, there are two bits of background information you need to get your head around.

Firstly, you may remember from my previous books that a *haiku* is a traditional form of Japanese poetry. I kinda dig 'em. Rather that following a rhyming pattern, *haiku* have a structure built around the number of syllables in the words in each line. The most common form of *haiku* is 3 lines long and has a syllabic structure 5-7-5.

Here's an example from Bashō Matsuo (1644–1694), considered the greatest *haiku* poet ever:

An old silent pond ...
A frog jumps into the pond,
splash! Silence again**

The second thing you need to know is that in the United States, the date that we Aussies refer to as '14 March' is written 'March 14'. In shorthand, this is sometimes abbreviated to just 3.14, as March is the third month of the year. But there's extra sauce with this because 3.14 also happens to be the most beautiful of all numbers, pi, approximated to two decimal places. So naturally, we maths nerds think of 14 March as 'pi approximation day'.

To celebrate this particularly geeky day in 2019, Twitter featured a series of *haiku* about pi, or 'pi-kus'.

* Stay tuned for more jaunty Japanese poetry forms on page 173.

** Obviously the original poem was in Japanese ...

I like this one by A. Chambert-Loir:

3 point one 4 one
Five nine two 6 5 3 5
Eight nine. And so on.

And this one by someone anonymous:

Just one Greek letter,
That sparks joy to math lovers,
Transcendentally.

And well done Toby Bailey who used the fraction approximation of π, 355/113, to give us this bad boy:

3 hundred and fif-
ty 5 divided by one
hundred and thirteen

Fractilicious

One thing I love about writing popular mathematics books is that I can, hopefully, take you across the threshold from looking at something that seems ugly, perhaps terrifying, and bring you out the other side with a feeling of calm and understanding. Take for instance continued fractions.

When you first see something like this:

$$2 + \cfrac{1}{1 + \cfrac{1}{4 + \cfrac{1}{3}}}$$

I can understand that you might feel a little bit queasy. But as is so often the case in my books, my advice is to take a deep breath, feel your heartbeat return to normal, take my hand and let's do this.

At the very bottom of this series of fractions you can see the fraction $4 + 1/3$.

We know how to simplify this into just one fraction:

$4 + 1/3 = 12/3 + 1/3 = 13/3$

You might remember from around 5th grade that we can just quickly calculate $4 \times 3 + 1 = 13$ and write this over the denominator of 3.

So our initial continued fraction becomes:

$$2 + \cfrac{1}{1 + \cfrac{1}{13/3}}$$

And again going back to primary school — and I'm saying this not to make you feel silly if you can't remember what to do, but to explain to my 'more mature' readers why it might be hard to find these tricks in the old memory! — we can write:

$$\cfrac{1}{13/3}$$

... simply as $3/13$. So our continued fraction now reads:

$$2 + \cfrac{1}{1 + \cfrac{3}{13}}$$

And continuing this way, we get:

$$2 + \cfrac{13}{16} = \cfrac{45}{16}$$

I've done the last few steps here, confident you can see what is going on. If not, go back and try it again, doing the same for the $1 + 3/13$ as we did for $4 + 1/3$ earlier.

The brutal mess we were originally confronted with, namely:

$$2 + \cfrac{1}{1 + \cfrac{1}{4 + \cfrac{1}{3}}}$$

... is just another way of writing the much more palatable fraction $45/16$.

How do we go back the other way? Given $45/16$, how do we create the continued fraction:

$$2 + \cfrac{1}{1 + \cfrac{1}{4 + \cfrac{1}{3}}}$$

Well, watch this. $45/16 = 2 + 13/16$. We can see this because $2 = 32/16$ and once we take 32 of the 16ths away from $45/16$ we have $13/16$ left over.

Now $13/16$ is less than 1 but we can flip it over and write $13/16 = 1/(16/13)$.

So $45/16 =$

$$2 + \cfrac{13}{16} = 2 + \cfrac{1}{16/13}$$

And $16/13 = 1 + 3/13$

So:

$$\cfrac{45}{16} = 2 + \cfrac{13}{16} = 2 + \cfrac{1}{16/13}$$

$$= 2 + \cfrac{1}{1 + \cfrac{3}{13}}$$

And doing the same to the $3/13$ as we did to the $13/16$ we further get:

$$2 + \cfrac{1}{1 + \cfrac{3}{13}}$$

$$= 2 + \cfrac{1}{1 + \cfrac{1}{13/3}}$$

$$= 2 + \cfrac{1}{1 + \cfrac{1}{4 + \cfrac{1}{3}}}$$

And because the final fraction in the chain, the $1/3$ has a 1 as its numerator, we are done.

When we write a number in continued fraction form with all numerators 1 we can use this shorthand notation, $45/16 = [2; 1, 4, 3]$ where the 1, 4 and 3 are the numbers under each fraction line (which is actually called a 'vinculum' if you're looking to add another awesome maths word to your vocabulary).

Here's a gorgeous diagram that also helps illustrate what we've done. We are examining $45/16$, so start by drawing a rectangle with sides … you guessed it … 45 and 16:

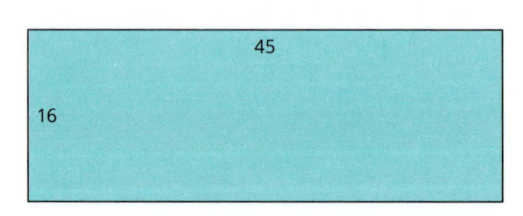

Now try to fit as many squares of 16 × 16 into the rectangle as you can.

You'll find you can fit 2 of them and you'll have a rectangle of 13 × 16 left over.

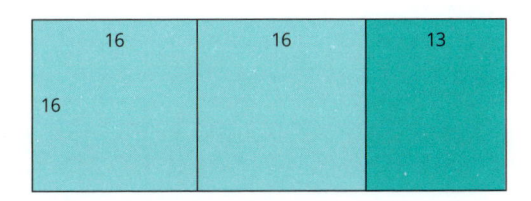

This diagram summarises the line of arithmetic you saw earlier, namely $45/16 = 2 + 13/16$.

If you can't see this equation mirrored in the diagram please take another moment to look at it again until it becomes clear.

Now do the same for the rectangle we have remaining. How many 13 × 13 squares can we fit into this 13 × 16 rectangle? You should get 1 and had a rectangle of 3 × 13 left over.

Can you work out which line of arithmetic this represents?

Well, we have effectively broken $16/13$ into $1 + 3/13$ so this diagram is showing us that:

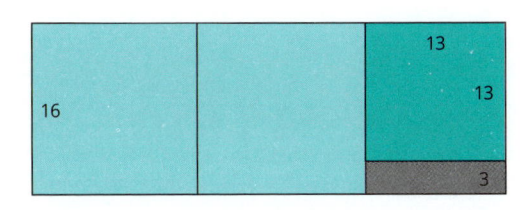

$$\frac{45}{16} = 2 + \cfrac{1}{1 + \cfrac{1}{13/3}}$$

Now ask yourself what should we do with this 3 × 13 rectangle. If you're thinking 'fill it with as many 3 × 3 squares as we can' well it's gold star time for you, my friend.

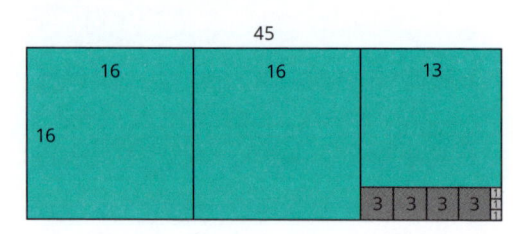

A big metaphorical thumbs up emoji from me to you. And once we've crammed four 3 × 3 squares in, the remaining rectangle is now 1 × 3. It can be filled entirely with 1 × 1 squares and we are done.

So looking at this rectangle and understanding the relationship between the geometry and the mathematics, the fact that filling a rectangle with squares and manipulating fractions is, in this case, exactly the same thing, you should be able to see how this diagram tells us that $45/16$ =

$$2 + \cfrac{1}{1 + \cfrac{1}{4 + \cfrac{1}{3}}}$$

I found this excellently explained on the website of Dr Ron Knott and he credits another maths guru Clarke Kimberling. Brilliant stuff, guys.

And would you really be surprised to know that that most gorgeous of numbers, pi, has some really cool continued fraction expressions? Expressed with numerators all of value 1 we get, π = [3; 7, 15, 1, 292, 1, 1, 1, 2, 1, 3, 1, ...]. Because pi is irrational it can never be written as an exact fraction. So the continued fraction representation of pi must go on forever.

Does that make sense? If we had a finite continued fraction, in the same way we started this chapter by working back from the continued fraction to get $^{45}/_{16}$, we could work back from π's finite continued fraction and get an expression for π as a fraction. We know that can't happen so the continued fraction expression of pi, like its decimal expansion, must be infinite.

Let's take a look at the continued fraction for π written out in its full fractional glory.

$$\pi = 3 + \cfrac{1}{7 + \cfrac{1}{15 + \cfrac{1}{1 + \cfrac{1}{292 + \cfrac{1}{1 + \cfrac{1}{1 + \cfrac{1}{1 + \cfrac{1}{2 + \cfrac{1}{1 + \cfrac{1}{3 + \cfrac{1}{1 + ...}}}}}}}}}}}}$$

Now if you're willing to relax the requirement that the numerators all be 1 ... well, turn the page my friend.

This allows for some truly gorgeous ways of writing π as a continued fraction. There are also several generalised continued fractions for our friend π which have a perfectly regular structure. Try these on for size:

$$\pi = \cfrac{4}{1 + \cfrac{1^2}{2 + \cfrac{3^2}{2 + \cfrac{5^2}{2 + \cfrac{7^2}{2 + \cfrac{9^2}{2 + \dots}}}}}}$$

$$\pi = \cfrac{4}{1 + \cfrac{1^2}{3 + \cfrac{2^2}{5 + \cfrac{3^2}{7 + \cfrac{4^2}{9 + \dots}}}}}$$

$$3 + \cfrac{1^2}{6 + \cfrac{3^2}{6 + \cfrac{5^2}{6 + \cfrac{7^2}{6 + \cfrac{9^2}{6 + \dots}}}}}$$

$$\pi = 2 + \cfrac{2}{1 + \cfrac{1}{\frac{1}{2} + \cfrac{1}{\frac{1}{3} + \cfrac{1}{\frac{1}{4} + \dots}}}} = 2 + \cfrac{2}{1 + \cfrac{1 \times 2}{1 + \cfrac{2 \times 3}{1 + \cfrac{3 \times 4}{1 + \dots}}}}$$

Nice.

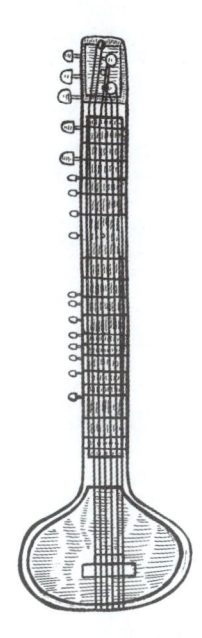

Steady on, sitar

The amazing Indian instrument the sitar was invented sometime in the 13th century but became best known in the West when the rock bands of the 1960s and 70s (the Beatles, Rolling Stones, et al) started experimenting with it in their music.

It's certainly a versatile instrument and can have 18, 19, 20 or 21 strings. The sitar player only plays 6 or 7 of them (the ones that run over the fretboard) while the others are known as 'sympathetic'. They run under the frets and resonate with their compadres above. Nice teamwork.

The famous Indian sitar player Ravi Shankar once played at New York's Madison Square Garden. After a few minutes on stage Ravi and his band received a huge round of applause. He turned to the crowd and said, 'Thank you. If you appreciate the tuning so much, I hope you'll enjoy the playing more!'

Hey, if things with strings are your, ah, thing, check out the stories on the theorbo and the cello on pages 50 and 277, respectively.

'The advancement and perfection of mathematics are intimately connected with the prosperity of the state.'

—Napoleon Bonaparte

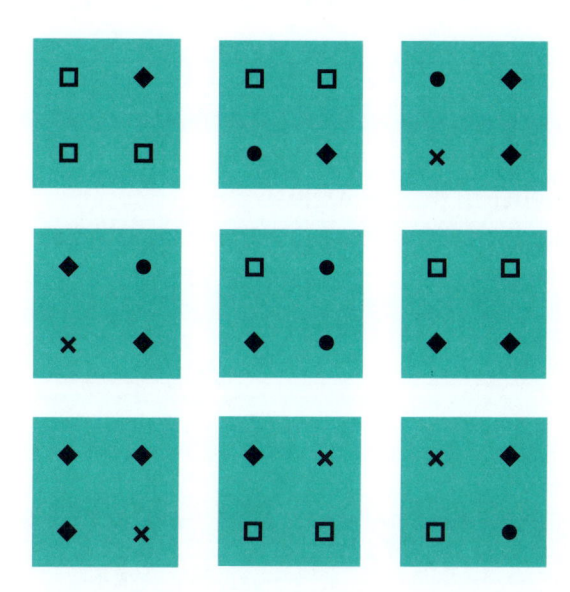

Tile on

Back to the bathroom in *Numberland*.

And onto the next section of our imaginary tiling job.

I've got to level with you. I'm probably going to run short on HILARIOUS introductory banter about tiling our bathroom in *Numberland*, so I'll cut to the chase from here.

There's still more tiling to be done, so don't put your mortarboard down just yet. Remember to place the 9 tiles into a 3 × 3 grid so that adjacent tiles show the same symbol along their touching edges.

The tiles are still all up the right way, but enjoy this hint because it won't last. Also enjoy the top left hand tile already being in the correct position. It's the last time I'll be so generous.

Grab the grout and get into it!

As always, you can head over to my website to download a cut-uppable copy of this puzzle.

Point your browser to adamspencer.com.au

105

Phenomenal Ford circles

This beautiful diagram illustrates mathematical objects known as Ford circles.

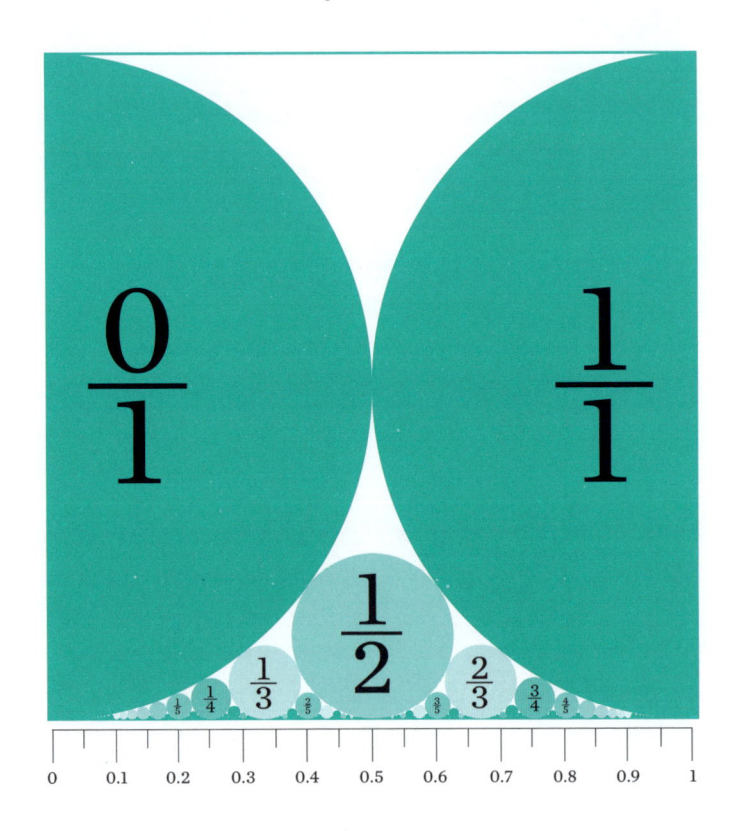

Ford circles are an infinity of circles, sitting within what we call a 'unit square' — a square of side one that in this case sits on the xy-plane with corners at (0, 0), (0, 1), (1, 0) and (1, 1).

Each circle's size and position are related by the following rule. The reduced fraction p/q on the number line from 0 to 1 generates a circle with its centre at the point with coordinates $p/q, 1/(2q^2)$ and radius $1/(2q^2)$.

What's that? 'Ugh, Adam, I'm feeling a little queasy, can you go over that again?'

Sure. Pick a fraction between 0 and 1. Okay, let's go for $2/3$.

If $p/q = 2/3$ then our rule generates a circle with its centre at the point $(p/q, 1/2q^2) = (2/3, 1/18)$ and radius $1/18$.

So this circle generated by $(2/3)$ which we label C[2, 3] looks like the one on the right.

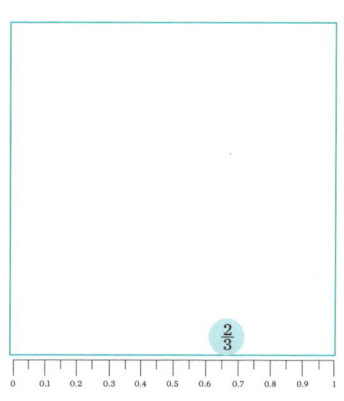

Similarly, the circle C[1, 4] sits at $(1/4, 1/32)$ and has radius $1/32$.

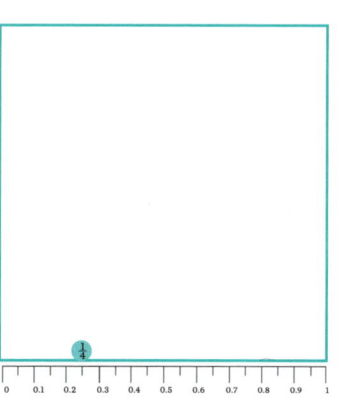

You should be able to see that because every Ford circle has the y value of its centre and its radius both equal to $1/2q^2$ they will always sit on the number line.

But what's beautiful about Ford circles is that none of them overlap. Any two Ford circles either 'kiss' each other at a single tangent point or they are 'disjoint' — a mathematician's way of saying, 'I like, want to have totally nothing to do with you.'

Even more awesomely, we can work out which circles are tangent to each other using another beautiful piece of mathematics called a 'Stern-Brocot tree' which will loop in our new friends we met earlier, the continued fractions.

What a crazy ride this all is. Let's start with the name, because that in itself is a story from which we can learn something. A Stern-Brocot tree is so named because it was discovered independently in 1858 by German number theorist Moritz Stern and then again in 1861 by French chronographer Achille Brocot who was designing gears for clocks.

Back before the internet (yes kids, there was such a time — we read books and ran around outside just in case you're wondering) it was common for people in different parts of the world to discover mathematical truths separately, completely unaware someone else had already stumbled onto the same thing. In the worst cases this led to massive arguments between mathematicians as to who deserved credit for something and who was just a cheap rip-off artist. In more chilled cases like this, we just named the tree after both of them and everyone seemed to be cool with that.

Anyway, a Stern-Brocot tree is a way of ordering fractions that looks a little something like this:

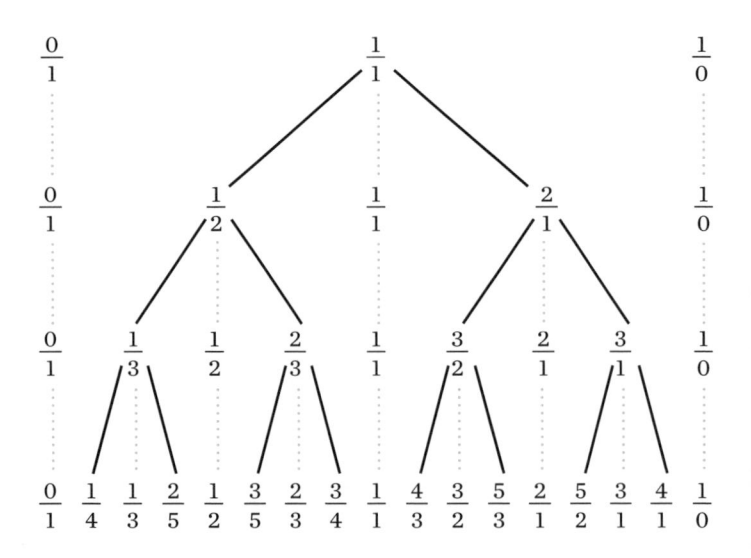

$$\frac{0}{1} \quad \frac{1}{1} \quad \frac{1}{0}$$

$$\frac{0}{1} \quad \frac{1}{2} \quad \frac{1}{1} \quad \frac{2}{1} \quad \frac{1}{0}$$

$$\frac{0}{1} \quad \frac{1}{3} \quad \frac{1}{2} \quad \frac{2}{3} \quad \frac{1}{1} \quad \frac{3}{2} \quad \frac{2}{1} \quad \frac{3}{1} \quad \frac{1}{0}$$

$$\frac{0}{1} \quad \frac{1}{4} \quad \frac{1}{3} \quad \frac{2}{5} \quad \frac{1}{2} \quad \frac{3}{5} \quad \frac{2}{3} \quad \frac{3}{4} \quad \frac{1}{1} \quad \frac{4}{3} \quad \frac{3}{2} \quad \frac{5}{3} \quad \frac{2}{1} \quad \frac{5}{2} \quad \frac{3}{1} \quad \frac{4}{1} \quad \frac{1}{0}$$

If you're feeling queasy again, take a deep breath and remember you felt this way a few minutes ago and it passed.

Ignore the right-hand side of the tree and focus on the fractions between 0 and 1. On the first row of the tree we have just $0/1$ and $1/1$. The second row contains $0/1$, $1/1$ and $1/2$. The third row features $0/1$, $1/1$, $1/2$, $1/3$ and $2/3$ and by row 4 we have $0/1$, $1/1$, $1/2$, $1/3$, $2/3$, $1/4$, $2/4$ (which is $1/2$), $3/4$, $2/5$ and $3/5$.

So you can sort of see that we seem to be pretty effectively picking up all fractions p/q here with $p \leq q$.

When we look at fractions joined by branches of a Stern-Brocot tree, for example 1/3 and 2/5, we say that 2/5 is the 'child' of 1/3 and that 1/3 is the 'parent' of 2/5. In fact, 1/3 has two children. The child on the left, 1/4 < 1/3, and the child on the right, 2/5 > 1/3. It gets a bit weird if you think that each parent has one child younger than it and one child older than its own parent so maybe park that thought for a moment. But as we continue down the tree, this relationship between parents and children continues — the child on the left is less than the parent, the child on the right is greater than the parent.

Further, looking at the 4th row, you see that even though 1/3 and 2/3 were revealed as children to 1/2 in the row above, they still sit in the order with 1/3 on the left of 1/2 and 2/3 on the right.

This should further make you confident we will pick up all fractions between 0 and 1 eventually on this half of the tree.

So each fraction on the tree has one parent and two children, and the relationship between parents and children comes down to ... you guessed it ... well, I gave you a pretty big clue earlier ... continued fractions.

It's a bit messy but let's have a crack.

Take our friend 2/3. We can write it as a continued fraction a couple of different ways.

$$\frac{2}{3} = \frac{1}{\frac{3}{2}} = \frac{1}{1 + \frac{1}{2}}$$

Or, continuing the process ...

$$\frac{1}{1 + \frac{1}{1 + \frac{1}{1}}}$$

We can write 2/3 using the notation [0; 1, 2] = [0; 1, 1, 1]. The 0 here means that the term 2/3 is between 0 and 1. If our original fraction was, say 4 2/3 we would write it as [4; 1, 2] = [4; 1, 1, 1]

If you try a few more examples of continued fractions you should see that any time the bracket notation of the fraction ends with a 1, you can shorten the brackets by one term and add the 1 to the second last term. Here [0; 1, 1, 1] = [0; 1, 1 + 1] = [0; 1, 2].

More generally, in fancy-pants maths speak, the continued fractions can be written two ways because:

$$[a_0; a_1, a_2, \ldots, a_{k-1}, 1] = [a_0; a_1, a_2, \ldots, a_{k-1} + 1]$$

For any fraction in the Stern-Brocot tree, write it these two separate ways. To get its children, just add 1 to the last term in each bracket.

Here the same continued fraction written two different ways, 2/3 = [0; 1, 1, 1] = [0; 1, 2] generates two children:

[0; 1, 1, 1] generates [0;1,1,1+1] = [0; 1, 1, 2], and [0; 1, 2] generates [0; 1, 2 + 1] = [0; 1, 3].

Why not quickly bash these out and check that:

[0; 1, 1, 2] = 3/5

and

[0; 1, 3] = 3/4.

Wow, we are getting this. I think you can give a little fist pump for that. I certainly did when I was first working through the Stern-Brocot tree.

Given that we now have 3/5 as a continued fraction, try to work out its two children the same way we derived it from 2/3.

You should get 4/7 and 5/8. If you're really hungry, why not try to derive all the children that would feature on the next row of the tree? We will see that next row very soon so you'll be able to check easily if you've got them correct.

One other really cute thing while we are here, for a parent–child couple where the smaller fraction is p_1/q_1, and the larger fraction is p_2/q_2 we have:

$$p_2 q_1 - p_1 q_2 = \pm 1$$

Check that this works for the values on the tree so far, and try to work out what determines if the answer is 1 or −1.

Do you remember that this all started way back when we were drawing Ford circles? How did we find ourselves crunching out continued fractions on the branches of the Stern-Brocot tree?

It turns out that every Ford circle touches an infinite number of other Ford circles. And ... get this ... the circles it touches ... can be read off the Stern-Brocot tree!

If we add another row to the tree we get:

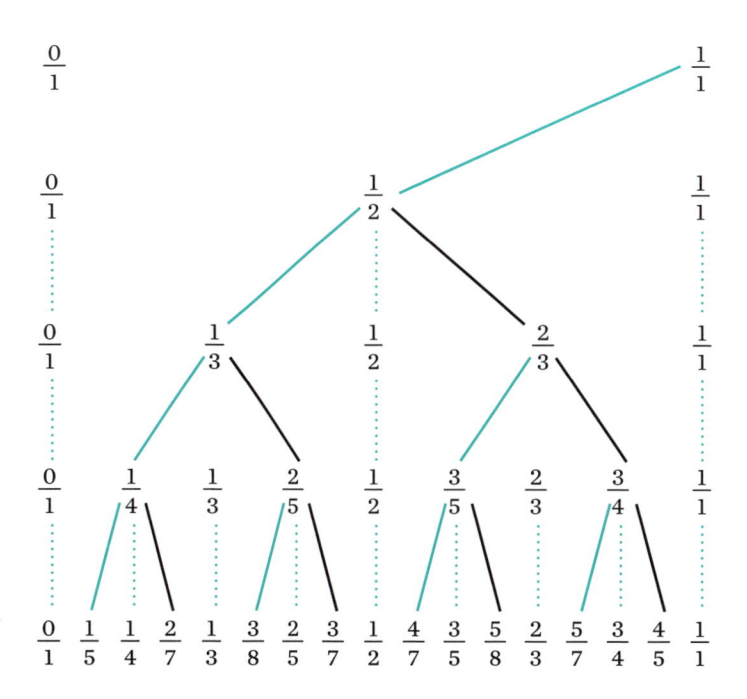

And let's also look again at the Ford circles.

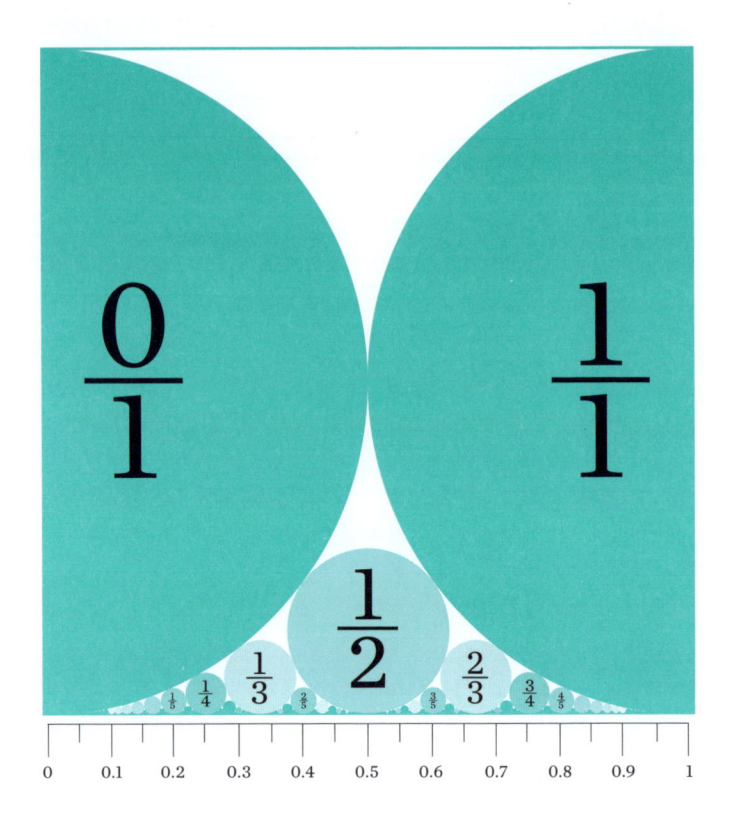

If you look at the children of 1/2, namely 1/3 and 2/3, you can see on the diagram of Ford circles that the circles C[1, 2], C[1, 3] and C[2, 3] touch. Then on the next line where 1/2 sits in between 2/5 and 3/5, the corresponding circles touch C[1, 2]. Go down another row and you'll see that the circle C[1, 2] touches C[3, 7] and C[4, 7]. And so on … forever.

Isn't that just gorgeous?

And if you're wondering, 'Hey Adam, of the entire square that clearly has area 1, how much of the space is taken up by the infinity of Ford circles?' I'm glad you asked.

Some truly awesome mathematics that lies just outside the reach of this book shows the circles cover about 0.872284041 of the square.

> One final question for the super-curious. Look at examples where C[p,q] touches two larger circles C[a,b] and C[c,d]. What relationship can you see between p, q, a, b, c and d?
>
> Answer as always, in the back of the book.

A whale of a time

Blue whales are big.

You might already be familiar with that factoid. But how about this one? At birth, the calves are up to 8 metres long and weigh 4000 kilograms.

An adult blue whale weighs an enormous 140,000 kilograms — give or take a kilogram or two.

But here's what blows my mind.

A blue whale's mouth is *so* big and stretchy, it can contain an amount of water that weighs *more* than the whale itself.

At 4° Celsius, which is when freshwater has its maximum density, a litre of freshwater weighs 1 kilogram. So how much would you expect 1 litre of saltwater at 25° Celsius to weigh?

A litre of seawater, on average, weighs 1.025 kilograms.

And a kilogram of seawater has, on average, about 35 grams of salt in it.

ADAM SPENCER

Snakes and adders

9	×	8	×	3
+	4	+	4	+
7	+	3	−	2
+	8	×	6	=
6	×	5	=	2

9	−	6	+	2
+	5	×	7	−
9	×	9	+	2
×	9	+	2	=
2	+	3	=	3

5	+	8	+	2
×	4	+	4	+
7	×	1	×	6
+	9	−	3	=
7	+	1	=	4

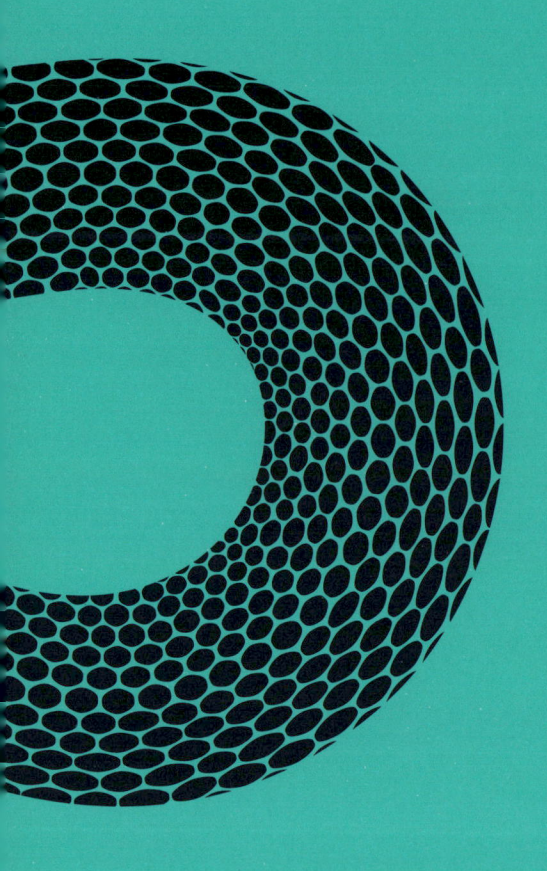

Remember our snakes from earlier?

Here's a refresher: in these grids, you can make paths from the top-left corner to the bottom-right by moving between adjacent squares.

For each grid, see if you can find 3 different paths that all trace out correct equations.

These get messy, so probably best to write out or photocopy a few of each grid and have your first bash with a pencil. You can also head to my website to download a copy to scribble on.

As always in *Numberland*, order of operations applies.

YOU ARE HERE

114

NUMBERLAND

→
8	+	6	+	2
×	4	−	5	−
2	+	3	+	2
+	5	+	6	=
1	×	2	=	5

→
4	×	3	+	8
+	1	−	9	+
8	×	9	×	2
×	3	×	8	=
3	×	4	=	6

→
8	+	3	+	7
+	9	×	6	+
8	×	6	×	8
−	3	+	5	=
2	×	9	=	7

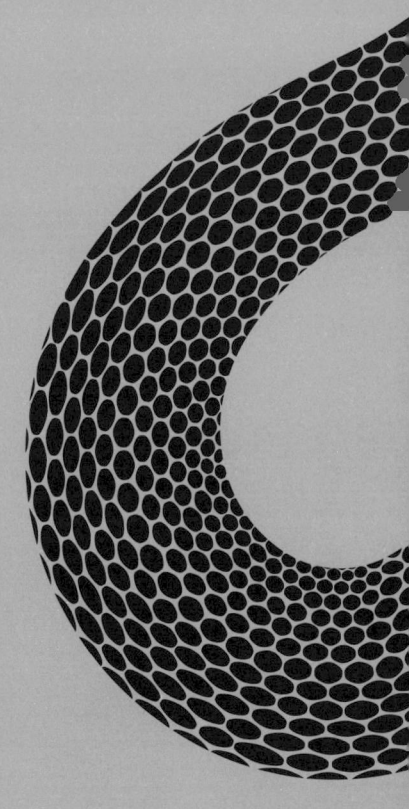

#PandasEatMilk
DudsAndSkittles

YOU ARE HERE

115

ADAM SPENCER

Here's something I hope you never need to know …

... and I only know because I saw it on the front page of the paper once. But the phrase 'one eighty-seven' is prison slang for murder.

It derives from Section 187 of the *California Penal Code* which reads 'Murder is the unlawful killing of a human being, or a foetus, with malice aforethought'.

10:09:36 — happy faces

One thing about this awesome journey we are all on called life is that you never stop learning. Clichéd yes, cheesy yeah a little bit, but rolled gold 100% bona fide true.

And the great thing about learning is it can happen at any time and come from any source.

So when my 10-year-old daughter Olivia mentioned the following fact to me, I was thrilled on many levels. The numerical nerdiness is beautiful but it was outshone by the fact that my little 'un* was beaming as she told Dad something about a number he'd never heard of before.

'Hey Dad, did you know that in every watch or clock commercial, the time is always 10.10?'

She's right!

But why is it so, as the great Julius Sumner Miller used to ponder.

It's unlikely to surprise you that there is more than one urban myth surrounding this little fact.

Nope, it's not to memorialise the time at which JFK/MLK/Abe Lincoln were shot (12.30 pm CST, 6.01 pm and 10.15 pm respectively).

Nor is it a tribute to the lives lost when the atomic bombs were dropped in Japan — Hiroshima was bombed at 8.15 am and Nagasaki almost 3 hours later at 11.02 am, local time.

* I should mention that Olivia is now 11 and by the time this book is released will be 12 years old. No offence to any 10-year-old readers, but 12-year-olds take being 12 VERY seriously.

The truth is far less romantic and far more pragmatic, as is so often the case.

Grab a watch and have a look. Can you figure it out?

Well, when a watch reads 10.10, the hands aren't overlapping, for a start. The time is pretty symmetrical. If you've got any complications (such as a date or stopwatch), it's unlikely they're covered either. Oh, and note that all-important logo, usually under the 12, is not just clearly visible but beautifully framed by the hands.

Ladies and gentlemen, if you guessed it's all about the marketing, then you're absolutely on the money.

What about, say, 8.20? Wouldn't that achieve the same result?

Ah, my chronologically-curious friend, you're right. In fact, there was a time when 8.20 was the standard, however the advertising folk realised it made it look like the watch was frowning, so the frowny setting was turned upside down into a smile.

But this isn't the only numerical nod to aesthetics you'll find on a watch dial, of course ...

Watchmaker Timex are even more specific with their watches, setting their times at 10:09:36 precisely.

Latin time

If you spot a watchface which uses Roman numerals, there's a good chance that the 4, rather than be depicted with the usual IV, will be written as IIII.

If you're putting your money on aesthetics once again ... there's a decent chance you're on the money. But hold your horses: it's not that simple this time. There is a range of theories as to why this may be the case.

Some of my fellow number nerds may be aware that the numeral 4 was, very early on, often depicted as IIII. Likewise 9, rather than being written as IX, was often written VIIII. It's thought that the change came about as these forms were harder to misread or confuse.

But the Romans didn't invent mechanical clocks — these were created later on in Europe in the 13th century. Roman numerals were still in use and many were likely mounted in churches at a time when Latin was still the official Catholic language, so it makes sense that clock faces would reflect this. Doesn't help us much in our IIII vs IV debate, though.

Returning to ancient Rome, the sundial was very much in vogue as a way of telling the time. In fact, many *pocket* sundials have been unearthed by archaeologists

Big Ben in London eschews the aesthetic imperative and sticks with the good old IV. Chime on, Ben!

through the years! How cool is that? Anyway, these artifacts have revealed a mix of both styles, however — so no dice. Another theory suggests that some ancient Romans were fearful of offending Jupiter, whose name spelled in Latin was IVPPITER. Others suggest it could be a nod to more uneducated watch-watchers, since the 'additive' IIII is more self-explanatory than the 'subtractive' IV, yet others pin the answer on Louis XIV, France's *Roi Soleil* or 'Sun King'.

This theory suggests that, as Romans might have been wary of offending Jupiter, French watchmakers might have been worried about offending their monarch — him being the embodiment of God on Earth and all that.

But let's face it, a lot of this speculation is more than likely the product of watch nerds with too much time on their hands (so many puns! Face it! Time! Hands! … so little time …) and auctioneers upping the price of watches for sale … the truth is once again likely far more pragmatic.

Here goes. Let's face it: IV and VI can be easily misread since, at their position on the watch, you're reading right to left. Also, sometimes, watch designers think IIII looks better. And sometimes, they think IV looks better.

The most expensive watch ever sold at auction was the Patek Philippe Henry Graves Supercomplication, which fetched US$23.98 million (23,237,000 CHF) in Geneva on 11 November 2014.

Sorry to say it didn't use Roman numerals, and the time in the pictures online is completely random. Oh, and just to — er — complicate things further, its analog display shows 24-hour increments on the face …

May I Have A Large Container Of Coffee?

Given how much caffeine I get from copious cans of a certain cold caffeinated beverage, that's probably not a great idea.

But it is another nice addition to our nerdy mnemonics in *Numberland*. Can you figure out what it is?

It's a device to help you remember pi to 7 decimal places. The number of characters in each word gives the digits: May (3), I (1) have (4) a (1) large (5) container (9) of (2) coffee (6)? π = 3.1415926

And if you want 14 decimal places, simply remember 'Now I Need A Drink, Alcoholic Of Course, After The Heavy Lectures Involving Quantum Mechanics'. π = 3.14159265358979

I've got to be honest, you may be better off just trying to memorise the digits ... but it's your call.

The loneliest number

Didja know that in Switzerland, it is illegal to keep just one, single guinea pig?

That might sound a bit ridiculous but the idea is noble — guinea pigs are social creatures and require interaction with fellow cute bundles of fluff to ensure their wellbeing. Keeping one is considered animal cruelty. In fact, to keep you on the right side of the law, you can even rent a guinea pig to keep your pet company should one of your dynamic duo die.

As far as I know it is still legal to have only one child in Switzerland ...

I haven't looked into it, but I doubt very much that similar laws apply in Peru, where the guinea pig — or *cuye* in Spanish — is considered a tasty delicacy.

Now before you go 'ewwww — that's gross!' you should also know that there's some research to suggest that they are a more environmentally sound alternative to cattle ... and one review I found online described the flesh as 'delicious, very tender and hard to compare to anything else'.

But before you start eyeing off the family pet, the same article's author, Alastair Bland of NPR, also described the meat as 'sinewy ... dry and sparse'. I suppose it's a matter of taste ... get it ... *taste*! Okay, tough crowd.

Hey, one man's guinea pig kebab is another man's pet, I guess ...

The odds of getting a triple-yolk egg are 1 in 25,000,000

That's according to the British Egg Information Service. The odds of a double-yolker are 1 in 1000.

So Dubbo-based Aussie Diane Wheeler was pretty chuffed when she cracked one while baking New Year's cakes in late 2018.

What about the odds of a green egg[*], like the one illustrated here? Dunno, but I wouldn't eat it if you find one.

[*] Dr Seuss' *Green Eggs and Ham* however is *not* to be avoided. The tale of the narrator and his nemesis, Sam-I-Am, uses only 50 different words, has sold over 8 million copies, inspired rapper Will.i.am's choice of name and is regularly voted one of the best children's books of all time.

Happy (nerdy) birthday, Jim!

I've not met Jim Simons, but he seems like a fun guy.

He's a mathematician who often tweets maths stuff. His Twitter profile spells out that while he is the mathematician, bridge player and dog lover, Jim Simons, he is *not* the *other* mathematician, string theorist and billionaire hedge fund guru, Jim Simons.

You won't be surprised to know that mathematicians celebrate their birthdays in pretty nerdy ways. After all, it's about numbers! So on 26 February 2019, Jimmy boy posted this riddle.

'It's my birthday. My age is the product of 3 distinct primes, as it was 4 years ago. How old am I?'

Why not give it a shot?

As I said, we maths nerds like doing this on our birthdays. So riddle me this. This year, for only the second and last time in my life, my age went from being a square to the double of a square.

How old did I turn?

The answer to both of these party crackers is at the back of the book.

Glacial pace

While we describe something moving at a 'glacial' pace to mean it is going incredibly slowly, it seems even among these giant bodies of ice, some absolutely fly, while some move ... I guess, well, glacially ...

Jakobshavn Isbræ (or 'Jacob's Glacier' to you non-Greenlandic speakers), for instance, moves at quite a clip.

Researchers from the University of Washington and the German Space Agency observed the glacier moving at a pace of more than 17 kilometres per year in the summer of 2012 — or over 46 metres per day.

Although these speeds slow down again in winter, the increase in pace over the previous few years averaged nearly 3 times anything recorded in the previous 90 which, you've probably guessed, is causing some alarm.

The University of Washington's Ian Joughlin says, 'We know that from 2000 to 2010 this glacier alone increased sea level by about 1 millimetre. With the additional speed it likely will contribute a bit more than this over the next decade.'

The point at which a glacier meets the sea and breaks off icebergs is known as the 'calving front'. With the general warming trend in the Arctic, these calving fronts are retreating further each year — between 2012 and 2013 alone, Jakobshavn Isbræ retreated more than a kilometre further inland than previously observed in summer.

If the trend continues, Jakobshavn Isbræ's calving front could retreat some 50 kilometres upstream to the head of the fjord from which it flows.

And that pace is anything but glacial by glacial standards ...

Jakobshavn Isbræ has another historic claim to fame: it is widely believed to have produced the large iceberg which sank the *Titanic* in 1912.

Ever hear the one about the infinite number of math nerds?

An infinite number of math nerds walk into a bar.

The first one orders a beer.

The second one orders half a beer.

The third one orders a quarter of a beer.

The fourth one steps forward with a smile on his face and is about to order, when the barman says 'you're all idiots,' and pours two beers …

Which leads us to ...

The curious sport of Infinite Jumping.

In this sport, our hypothetical athlete is able to jump.

Now, you might suggest that many athletes share this ability, however stay with me. *Our* hypothetical athlete has one special distinction: she is able to jump *forever*. And you thought a cricket Test takes a while ...

Although superhuman, every time our jumpy friend, er, jumps, she gets a little more tired. As a result, every jump she makes goes *half* as far as the previous jump.

For her very first jump, she travels half a metre. The crowd goes wild!

On her second jump, she goes a quarter of a metre. The crowd is on the edge of its seat awaiting the third jump ... in which she travels half as far again.

Okay, I'll admit this probably works better as a brainteaser than an actual sport.

Anyway, on that note, how many jumps would it take for our athletic friend to travel a whole metre?

ADAM SPENCER

... or the one about the lottery for an infinite amount of cash?

$$$$$$$$

Okay, last infinite joke for the moment. Promise.

So, a mathematician organises a lottery ...

The prize? An infinite amount of money.

When the winning ticket is drawn and the absolutely ecstatic winner comes to claim her prize, the mathematician explains the mode of payment: 'One dollar now, ½ a dollar next week, ⅓ of a dollar the week after that ...'*

* If you're not ROFLing right now, perhaps you should search the phrase 'harmonic series' online.

```
wwwwwwwwwwwwwwwwwwwwwwwwwwwwww
wwwwwwwwwwwwwwwwwwwwwwwwwwwwww
wwwwwwwwwwwwwwwwwwwwwwwwwwwwww
wwwwwwwwwwwwwwwwwwwwwwwwwwwwww
wwwwwwwwwwwwwwwwwwwwwwwwwwwwww
wwwwwwwwwwwwwwwwwwwwwMwwwwwwww
wwwwwwwwwwwwwwwwwwwwwwwwwwwwww
wwwwwwwwwwwwwwwwwwwwwwwwwwwwww
wwwwwwwwwwwwwwwwwwwwwwwwwwwwww
wwwwwwwwwwwwwwwwwwwwwwwwwwwwww
wwwwwwwwwwwwwwwwwwwwwwwwwwwwww
wwwwwwwwwwwwwwwwwwwwwwwwwwwwww
wwwwwwwwwwwwwwwwwwwwwwwwwwwwww
wwwwwwwwwwwwwwwwwwwwwwwwwwwwww
wwwwwwwwwwwwwwwwwwwwwwwwwwwwww
wwwwwwwwwwwwwwwwwwwwwwwwwwwwww
wwwwwwwwwwwwwwwwwwwwwwwwwwwwww
wwwwwwwwwwwwwwwwwwwwwwwwwwwwww
wwwwwwwwwwwwwwwwwwwwwwwwwwwwww
wwwwwwwwwwwwwwwwwwwwwwwwwwwwww
wwwwwwwwwwwwwwwwwwwwwwwwwwwwww
wwwwwwwwwwwwwwwwwwwwwwwwwwwwww
wwwwwwwwwwwwwwwwwwwwwwwwwwwwww
wwwwwwwwwwwwwwwwwwwwwwwwwwwwww
wwwwwwwwwwwwwwwwwwwwwwwwwwwwww
wwwwwwwwwwwwwwwwwwwwwwwwwwwwww
```

Memebusters

Above is a bit of a meme that was going around recently.

According to the good folk of internetville, if you can find the one 'M' in the list of 779 'Ws', you're a genius.

Look, I need to level with you: I haven't researched this in depth enough to verify that particular claim. But, my eagle-eyed friend, I have my doubts.

Nevertheless, it's a bit of fun. Give it a go.

Elephants in *Numberland*

Here are a few runners-up of crime by the numbers.

*To help combat crime on New York subways, plain clothes police were given packs of cards bearing the mugshots of 50 of the best known repeat offender pickpockets. These thieves became known as **The Nifty-Fifty**.*

*In Denmark the **AK81** are a branch of the Hells Angels motorcycle gang that provide muscle for battles over the lucrative drug market. From what I've read online, they are dangerous dudes and pretty average human beings.*

*In 1982 and 83 a group of 6 teenagers from Milwaukee, Wisconsin hacked into computers at a series of high profile targets, including the Los Alamos National Laboratory (where nuclear weapons were pioneered during World War II). Using the phone prefix for their home town, they were known as the **414s**.*

*The **47 Ronin**, a band of 18th century revenge-seeking masterless samurai, are featured elsewhere in the book … and well worth a read.*

People often stop me in the street and ask, 'Hey Adam, what's the most interesting criminal gang you've ever heard about that use numbers in their names?'

No they don't. That's a lie. But a white lie, because I'm using it as a clunky entry into a story about a fascinating but very bad group of women from a while back.

If anyone were to ask me, of all the myriad numerical criminal gangs I've read about, my favourite would have to be the '40 Elephants Gang' in late 19th century London.

Running on a patch of turf in South London called Southwark, the 40 Elephants were a group of female English *badlasses* who counted themselves among the best shoplifters of the day.

These girls didn't just nick the occasional chocolate bar when the shopkeeper wasn't looking, they didn't swipe a magazine by saying, 'No, I already paid, you've just forgotten.' These women took it to the next level.*

Run by the ruthless Diamond Annie (cool name*), the women hid multiple fur coats, jewellery and other valuables inside the clothing they were already wearing. This was another reason they were called the 40 Elephants — as they waddled out of another store, their clothing bursting with ill-gotten gains, they thought and felt that they looked like, well, elephants.

NUMBER LAND

One of Annie's favourite stunts was to pull up to a fancy department store in a limousine, wearing an expensive fur coat she'd stolen from somewhere else before then loading up on vast amounts of top end gear.* One of the reasons they could do this was that women were given great privacy in changing areas of shops due to the conservative nature of the times.

They also pulled other cons like infiltrating wealthy families acting as serving maids before, you guessed it, nicking valuable stuff from the wealthy person's house. They'd sell off much of the stuff they swiped (again*).

It is thought the gang operated from at least 1873 until perhaps the 1950s ... maybe longer. At its height there were actually at least 70 Elephants in the 40 Elephants Gang and the children of previous gang members were known to join to keep the gang operating across the generations*.

In case you missed my liberal sprinkling of * during this article, please read the paragraph to the right.

* Shoplifting is really bad and you shouldn't do it. Like, totally, *don't do it*. I won't think you're cool or a badass. You will be doing something very wrong and you will get caught.

YOU ARE HERE

133

Think fast #1!

Try this one on for size, without thinking too hard about it. What's your first answer?

Say we're in a parallel universe in which I'm running the Comrades Marathon. I've been training for yonks and, after going flat out (and not passing out after the first kilometre, thank you very much), I finally pass the person in second place!

Phew! Go Adam!

Tell me, what place am I in now?

Harland Rose

Cute-as-a-button Harland Rose was born on 9 September 2018 — the same birthday as Colonel Harland Sanders of KFC fame.

That's no coincidence — little Ms Rose was the first baby born on this dubiously auspicious anniversary to be named 'Harland', thus winning a competition run by KFC to celebrate the Colonel's birthday.

The prize? A US$11,000 college scholarship which she will receive 18 years from now.*

The 3.66 kilogram bub was born to Anna Pilson and Decker Platt who say they hadn't initially considered the name 'Harland', but when their bundle of joy came into the world at 12 seconds past midnight on the 9th, they made up their minds. Little Harland will go by the nickname 'Harley', they say.

Hey, on the plus side, the rules of the stunt didn't require her to go by the name 'Colonel' ...

*Bear in mind that, given she won't receive the prize for some 18 years, the actual amount will be much larger. If the money is invested during that time with, say, a return of 8% compounded annually, the actual amount will wind up nearer US$44,000.

Kaaaaching, Harley.

Rock is for rookies

You know the drill ...

On the count of 3, you toss out a hand gesture against your opponent which represents either paper, scissors or rock. It's a game of luck rather than skill. Or is it?

Humans are terrible at being 'random', which means that there might be some level of predictability, after all. According to RPS veteran (hey, why not use a NA [nifty acronym] here?) and World Rock, Paper, Scissors Championships organiser Graham Walker, there are strategies that could give you an edge.

There are two main paths to RPS glory: either by eliminating an opponent's move, or by forcing them to make a predictable one. 'The key,' Walker says, 'is that it has to be done without them realising that you are manipulating them.' Game on.

Many veteran players have noticed that rookies, or irregular players, often throw down rock on their first move. No one's certain why, but the popular theory goes that it's considered strong, and it's a natural shape to form with your hand (especially since you're starting out with it in the initial 'throws'). So if this is the case, we can start to strategise how to win. If your opponent is a total noob, throw out paper — it could wrap up (*ha!*) the game quick smart.

Likewise, because rock is viewed by pros as being, well, predictable, they're also less likely to lead with it. If that's the case, your best bet would to be hedge with scissors (*C'mon!* Lame pun *streak*!) since you'll beat out paper, or at worst tie with your opponent's scissors throw.

But there are other RPS Jedi tricks you should also know before you attempt to settle, say, who should take the bins out with this ancient*, noble duel.

*The first recorded mention of the game we now know and love as RPS was by the Chinese Ming-dynasty writer Xie Zhaozhi in the 1600s, in the book *Wuzazu*. He wrote that the game dated back to some time between 206 BC–220 AD, the time of the Han dynasty.

The power of suggestion is incredibly important in human psychology and there are a number of techniques you can employ to mind-game your way to RPS victory. As you're readying for the game, try gesturing over and over with the move you want to 'will' your opponent to make next. The argument for this is that if your opponent is not paying careful attention, their subconscious might pick up the signal and take over. And there's some evidence to back that up. A study on decision making in RPS, published in the July 2011 edition of the *Proceedings of the Royal Society B*, suggests that players will often imitate their opponent's last moves.

That said, the pros have also observed that people will often try to overcome a loss by throwing whatever would've won their last round. See how you go, I guess ...

Of course, announcing outright what you're going to throw could also work — but only once. If you say you're going to throw a rock, your opponent will almost certainly presume you're lying.

But, finally, Walker offers this sneaky, low-blow strategy, to be deployed only if you're willing to sacrifice your soul for a guaranteed win.

Ready to give it a shot, Faust? Here goes.

'When you suggest a game with someone, make no mention of the number of rounds you are going to play. Play the first match and if you win, take it as a win. If you lose, without missing a beat start playing the "next" round on the assumption that it was a best 2 out of 3. No doubt you will hear protests from your opponent but stay firm and remind them that "no one plays best of one!"'

And yes, I *will* judge you if you try that one on.

In November 2018, English referee David McNamara was suspended for 3 weeks by the Football Association after having the Manchester City and Reading women's team captains play RPS to determine opening possession! McNamara thought it'd be a fun break from the traditional pre-game coin toss ... the FA thought otherwise.

Rep tiles

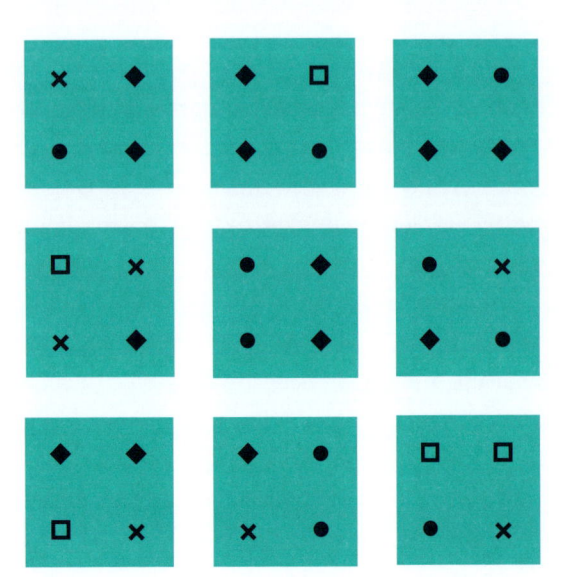

Think you're done tiling the bathroom? Think again, my friend!

Remember to place the 9 tiles into a 3 × 3 grid so that adjacent tiles show the same symbol along their touching edges.

The tiles are all facing the right way up.

Grab the grout and tile on!

As always, you can head over to my website to download a cut-uppable copy of this puzzle.

Point your browser to adamspencer.com.au

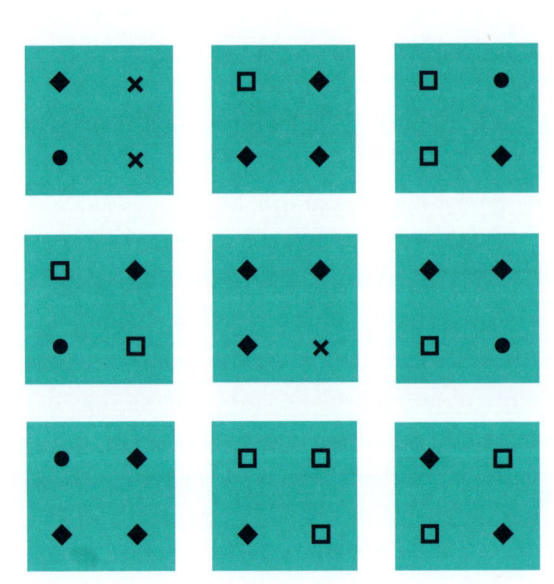

138

'All mathematicians share a sense of amazement over the infinite depth and mysterious beauty and usefulness of mathematics.'

—Martin Gardner

Australia's biggest polling booth station ...

Is in *Harry Potter*'s Gringott's Bank.

Yes, you read that correctly.

Now that our democracy sausages are downed and we're safe(ish) from campaigning for another few years, let's look at one of the more interesting statistics to arise from the 2019 Federal election.

At Australia House in London, Aussie ex-pats and tourists lined up to cast their vote in the Exhibition Hall which starred as the Gringott's Bank in the first *Harry Potter* film.

The booth anticipated over *15,000* Aussies to vote in the lead-up to 17 May (18 May back here in the Antipodes) by preparing some 45 cardboard booths staffed by a dedicated team of 28. A whopping 17,000 ballot papers made their way to Old Blighty from Australia, covering all 151 electorates.

Now, if only the leaders' debates were half as interesting as a *Harry Potter* film ...

Almost 17 million Australians were enrolled to vote in the 2019 election. They did so at 7000 different polling stations or, in a quarter of all cases, voted early.

For the 2016 election, the election funding rate for candidates, or at least those who secured at least 4% of the formal first preference vote, was 262.784 cents per vote.

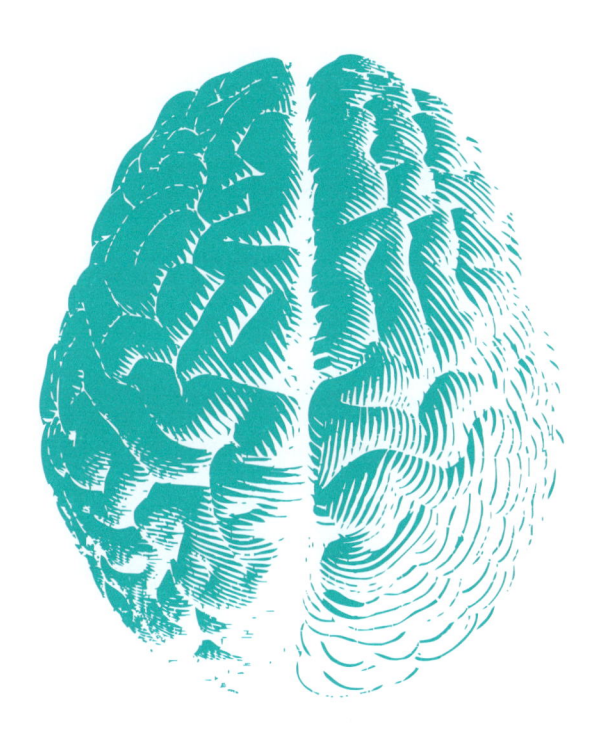

Think fast #2!

You have 5 seconds to answer this little brain-tingler.

Ready? Here goes.

David's mother has 3 children.
The first child was named April.
The second child was named May.
What was the third child's name?

141

Season 5, episode 4

Welcome to Casa di Soprano.

In episode 4 of season 5 in the TV classic *The Sopranos*, AJ is getting tutored for the SAT exams.

The sample question provided by his instructor is as follows: 'If a million zeroes can be written on the front and back of a sheet of paper, how many sheets of paper do you need for a googol of zeroes?'

Remember, a 'googol' is 1 followed by 100 zeroes.

Can you figure it out?

The Sopranos ran from 1999 to 2007 over 6 seasons totalling 86 episodes.

It did alright, winning 21 Emmy awards, 5 Golden Globes, and frequently topping lists of the 'greatest TV series ever made'.

'Mathemathics is the queen of the sciences.'

—Carl Friedrich Gauss

... who also said that 'the enchanting charms of this sublime science reveal only to those who have the courage to go deeply into it.'

So true, Carl!

Oops

Ever cringed after accidentally sending an important email to the wrong person?

Or sent your book to print with a glaring typo, for that matter? Nah, that would never happen, *cough*.

Spare a thought for one fat-fingered trader, quietly working away at Japanese broker Mizuho Securities Co. One day in 2005, our hapless friend accidentally listed 610,000 shares in their company at 1 yen apiece instead of the actual amount — 610,000 yen (almost $8000 Australian dollars).

Despite trying to cancel the mistake 3 times (I'd imagine in an increasingly panicked state), Mizuho was met with little sympathy from the Tokyo Stock Exchange, which offers no cancellations on any orders.

The mistake sent the Nikkei 225 Index down almost 2% and cost Mizuho US$225 million.

But this pales in comparison to the loss incurred in September 2014 when a trader accidentally pressed the wrong button and cancelled 42 separate stock sales totalling some 67.78 *trillion* yen. That's roughly US$617 *billion*.

London's *Evening Standard* newspaper observed that the mistake was 'thought to be the most extreme example of a trader in financial markets inputting hopelessly wrong figures while working under intense pressure'.

Quite.

Perhaps they should have raised the temperature in the office?

One study from researchers at Cornell University found that raising the temperature of an office by 5° Celsius can help reduce typos by 44% ... and increase productivity by some 150%.

Carry on!

Our trusty mathematical carrying method is good for numbers with just a few digits, but what happens when we need to multiply numbers with millions or billions of digits?

You and I might not be troubled by such things in our day-to-day lives, but computers are and they've not been happy about it. Why? Well, multiplying two numbers with 1 billion digits requires 1 billion squared, or 10^{18}, multiplications. This would take a computer decades!

For millennia we simply didn't know any other way of multiplying. It was laborious. It was a drag. Especially for the poor computers when they came into being. But in 1960, a 23-year-old Russian mathematician by the name of Anatoly Karatsuba took a class by fellow Russian mathematician Andrey Kolmogorov, one of the best maths brains of the 20th century. Kolmogorov stood at the lectern and argued that there was no known way of multiplying two numbers with n digits that required fewer than n^2 steps.

Suffice to say, 'El-Karat' (yes I'm making that nickname up, I presume most of his friends just called him Anatoly) disagreed and he got to work. Often when mathematicians sense something might be slightly off, it may take months, even years, to prove that nagging voice correct. Amazingly, just a week later, El-Karat proved Kolmo wrong. Here's how he did it.

Say you want to multiply two 200-digit numbers. Write them as $aX + b$ and $cX + d$, where $X = 10^{100}$. Then a is the first 100 digits of the first number and b is the last 100 digits of the first number, while c and d are the first and last 100 digits of the second number. We want to work out the product $(aX + b)(cX + d)$, which is $(ac)X^2 + (ad + bc)X + (bd)$.

The old-fashioned way would have had us doing all 4 multiplications ac, ad, bc, bd. El-Karat discovered that we can do it with just 3: ac, bd, and $(a + b)(c + d)$ ... and get our $(ad + bc)$ as $(a + b)(c + d) - ac - bd$. Instead of doing $200^2 = 40,000$ calculations, only 3 lots of 100^2, or 30,000 calculations are needed.

Basically, he broke up the digits and recombined them in a way that substituted a small number of additions and subtractions for a large number of multiplications. It was a game changer because repeated application of this trick meant that multiplying two n-digit numbers took more like $n^{(\log_2 3)}$, or about $n^{1.585}$ steps, as opposed to n^2 steps.

For even bigger numbers, the speed improvement becomes massive. Multiplying two 1024-digit numbers can be done with something in the order of $3^{10} = 59,049$ single digit multiplications, compared to over a million such operations under the traditional method!

But wait, there's more.

In 1971, mathematicians Arnold Schönhage and Volker Strassen came up with an even faster multiplication algorithm. Then in 2007, Swiss mathematician Martin Fürer came up with an algorithm called — you guessed it — Fürer's algorithm — which was faster still, but not all that practical. And there have been other advances since.

These later methods are a bit too complicated to explain in detail here. But just getting your head around the idea that computers multiply numbers differently to the way we did at school and understanding why that is important, is a pretty cool place to be at IMHO.

Then in 2019, Australian mathematics whiz David Harvey, along with his good mate Joris van der Hoeven, announced to the world that they had found a method that can multiply massive numbers using only $n \times \log n$ steps. Amazingly, the number of steps are no longer a square of n, but much closer to just n itself.

This would reduce our 1024-digit multiplication from earlier from over 1,000,000 steps to close to 1024 steps. For truly gigantic numbers the speed increase is even more impressive and important. Awesome work, guys. Such exciting times!*

*... geddit?
... Times!!!

'113' is the number for the Report-a-Spy hotline in South Korea

Put it on speed-dial if you're heading over there on holidays, I guess.

ADAM
SPENCER

The Dodo

Lewis Carroll's character in *Alice's Adventures in Wonderland* is the large but sadly now extinct flightless bird from the island of Mauritius.

Rumour has it Carroll based the character on himself, in part because he had a stammer and would accidentally introduce himself as 'Do-do-dodgson' (his real name was Charles Lutwidge Dodgson). Here are 3 nifty facts about ... the dodo.

The fearless dodo (*Raphus cucullatus*), a distant relative of the pigeon and the dove, dwelled merrily on the island paradise of Mauritius for hundreds of thousands of years (perhaps more) before the arrival of Dutch sailors in the 1500s. Within less than a century, following hunting and habitat destruction, the dodo joined our long and embarassing list of extinct species.

They were surprisingly big creatures — weighing in at up to 25 kilograms and standing an imposing 1 metre tall. As no complete specimens exist, we're guessing at their precise appearance, but scientists believe they were grey/brown with a grey bald head and a green beak. Their beaks were big, too — much bigger than their heads — giving them a slightly goofy look, but they weren't stupid. Just lucky to live on an island without enemies or predators. Until we arrived.

'The Dodo Diet' hasn't yet hit the bookshops ... but it probably should. A Dutch letter dating from 1631 suggests the dodo loved fresh fruit (mainly that which had fallen conveniently to the ground), seeds, nuts, bulbs and roots. Vegan *and* full o' fibre! It's also thought the dodo was partial to shellfish and small crabs as its beak was big and powerful. So balancing things out with a bit of sea protein. As delicious as a Dodo diet.

Illustration by John Tenniel

Iceland, baby

As is generally the case, the world was pretty interested in all things football during the 2018 World Cup.

But one country really put in the hard yards when it came to getting behind the national team.

In a gripping game, Iceland — population 300,000 in total, inclusive of men, women and children — drew 1-1 with Argentina.

To put that in perspective, Argentina's population is 44 million, with some 2,658,811 football players in total — 331,811 of whom are officially registered with 3377 football clubs. There are 37,161 Argentinian football officials, to boot.

That's right, Argentina has more officially registered players than the *entire* population of Iceland.

But that didn't stop the country getting behind *Strákarnir okkar* ('our boys').

It was estimated that of all the Icelanders watching TV at the time, a whopping 99.6% were tuned into the game.

Whose responsibilty was this, anyway?

Returning once more to the subject of financial oopsies ...

The Reserve Bank of Australia suffered the mother of all typos recently — and as we've seen, there's stiff competition.

Six months after it entered circulation, eagle-eyed spellers noticed that the new, polymer $50 note had a glaring (if very tiny) error — the word 'responsibility' has been spelled 'responsibilty'.

The mistake appears in the reproduction of the great Edith Cowan's maiden speech to Parliament in 1921, 'It is a great responsibility to be the only woman here, and I want to emphasise the necessity which exists for other women being here.' These were especially poignant words since Cowan was the first woman in the Australian Parliament, ever.

If you're thinking of holding onto one of the notes in case they become collectible in the future, bear in mind that the typo'd $50 was reproduced some 480 million times. In fact, there are 46 million already in circulation at time of print.

The RBA says that not only will the notes remain in circulation, they'll eventually release the remainder of the notes already printed. But I suspect they'll run a spell-checker over them before they go to reprint.

Thou shalt commit adultery ... NOT!

In our ongoing examination of typos in *Numberland*, I'd be remiss if I didn't give a nod to this little corker.

Whatever your beliefs, you're likely familiar with the Christian Ten Commandments set out in the Bible:*

1. Thou shalt have no other gods before me
2. Thou shall not make unto thee any graven image
3. Thou shalt not take the name of the Lord thy God in vain
4. Remember the sabbath day, to keep it holy
5. Honour thy father and thy mother
6. Thou shalt not commit murder
7. Thou shalt not commit adultery
8. Thou shalt not steal
9. Thou shalt not bear false witness against thy neighbour
10. Thou shalt not covet (basically anything)

It's fair to say that, all things considered, there's a few pretty decent Ideas to Live By in there — not least of all 'thou shalt not murder', or steal, for that matter.

It's not a commandment, but I seem to recall something about forgiveness in the Bible, too ... which is probably a good thing for the hapless 17th century typesetter who, in a 1631 edition of the King James Bible, omitted a fairly important word — 'not' — from the 7th commandment.

It read 'thou shalt commit adultery'. Oops.

*There are various wordings going around, but they all read a little something like this.

10^{99} is a duotrigintillion, or one million trigintillion

Just in case you were wondering ;-)

The future of transport?

Here's a story for anyone who's ever let a toddler go down the local slippery dip and wondered, hopefully just in the nick of time, 'Is this thing too steep for him/her to be on by him/her self?'

Spanish thrillseekers were forced to reconsider their life choices after queuing to take a turn on the newly-unveiled, 38-metre Estepona slide linking two streets on the Costa del Sol.

It was designed to allow quick travel from one street to another, thus sparing those in a hurry a boring 10-minute walk. But it turns out there's a reason slippery dips have not been widely used elsewhere as a mode of transport.

Social media users were quick to upload videos of 'commuters' hurtling at terrifying speeds from Point A to Point B before shooting off the end like Champagne corks out of a bottle.

Although the local authorities have pointed out that you're supposed to stay sitting up to slow your speed ... rather than lie down to turn yourself into a prostrate pedestrian projectile ... they've closed the slide indefinitely while they 'fine-tune' it.

Perhaps the final word should go to @AzulDebonisB who captioned photos of her grazed elbows on Twitter '*El tobogán de Estepona es una mierda, visto y comprobado. Me he tirado y me hice daño por todos lados, volé 2 metros y los policías se empezaron a reír.*' Which, my non-Spanish speaking friends, translates as 'The Estepona slide is a piece of crap. I went on it and got hurt all over. I flew 2 metres and the police started to laugh. I'm not putting up [a photo] of my bottom, which is worse'. Indeed.

Guinness World Records tells us that Action Park in New Jersey outside New York has the world's longest water slide. At 602 metres long it features 2048 metres of inflatable polyvinyl chloride and is made up of twenty 30-metre sections of inflatable material held together by 6,000,000 stitches.

It takes two generators to power the 15 engines that keep the slide inflated and 400 stakes to hold it down.

A trip from top to bottom takes around 90 seconds. Wow.

Learn as ewe go

Jule Ferry School in Crêts-en-Belledonne, a village in the Alps northeast of Grenoble, recently saw its student numbers fall from 266 to 261.

Unfortunately that meant that numbers fell below the threshold required by local authorities to keep the school open. But racing to the rescue was local herder Michel Girerd who took it upon himself to register some of his ewes. He appeared at the school with 50 of them for a ceremony which was attended by some 200 teachers, pupils and officials.

No word on how the students felt about their new classmates — among whom were Dolly, Shaun and (my favourite) Baa-bete — however local media reported children at the event waving signs which read 'we are not sheep'.

Flight control

You know that feeling when you keep putting something off?

Anyone who played that old aircraft controller game on the smartphone a few years back will tell you that it's a tricky business. I suspect you'll be unsurprised to learn that the real thing is even more complicated.

Los Angeles air traffic control is reliant on an internal clock which operates in a special way: it counts down from 4,294,967,295* once every millisecond in order to keep track of precise time.

What happens when it runs out? We are not meant to know because it's meant to be restarted before it gets too low.

But that wasn't the case on 14 September 2004 when, due to an 'administrative oversight' it hit zero and promptly shut the entire system down, leaving some 800 aircraft puttering about Southern California without air traffic control.

Staff wound up frantically trying to use mobile phones to contact aircraft to tell them to connect to a different control centre while no less than 10 planes flew closer to each other than regulations would ordinarily allow.

Fortunately nobody was injured, but thousands were inconvenienced as flights were subsequently delayed or cancelled for some time afterwards.

Moral of the story? Sometimes it's better not to put things off till the last minute …

---

* Why 4,294,967,295? Well, 2^{32} = 4,294,967,296 and, as a result, is the largest value you can store in a 32-bit operating system.

And in case you're wondering, 4,294,967,295 milliseconds is 49 days, 17 hours, 2 minutes … and 47.295 seconds.

Mayhew the Force be with you

Star Wars nerds shed a tear when on 30 April 2019 Peter Mayhew died at the age of 74.

The 221-centimetre Mayhew (7 foot 3 in old-speak) played the character of Chewbacca in 7 of the *Star Wars* instalments.

To commemorate the great man and to really touch the hearts of *Star Wars* loving readers, I thought I'd use the website wookietranslator.com to write the phrase 'Peter you will be missed. Well played, mate' into the Wookie* language.

Well, it came out as *'aguhwwgggghhh uughghhhgh wrrhwrwwhwuughguughhhghghghhhgh uuggghuughgu-ughhhghghghhhgh aaaaahnr aarrragghuuhw'*.

I thought I'd check again with the translator and it came out as *'uughguughhhghghghhhgh huurh huuguughghghuu-guughghg hnnnhrrhhh uughghhhgh huurh huuguughghg'*.

And a third time, when Wookie Translator interpreted 'Peter you will be missed. Well played, mate' as *'uugh-guughhhghghghhhgh huurh raaaaaahhghhuuguughghg hnnnhrrhhh huuguughghg uuh aaaaahnr'*.

I'd respectfully suggest that Wookie Translator isn't the most rigorous language site going around. Perhaps I should just leave it at this: Peter you will be missed.

Well played, mate!

* Officially spelled with two Es, but since the website name gets it wrong, let's roll with it ...

Operation!

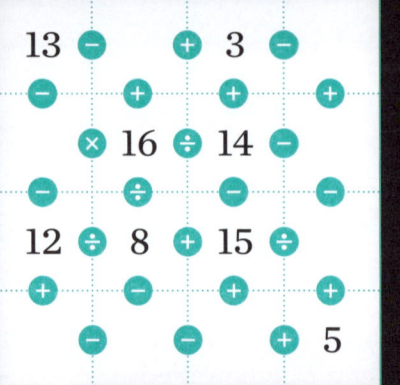

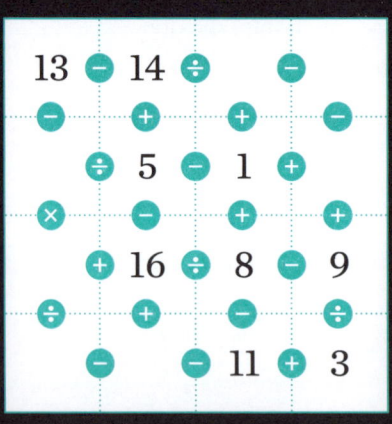

Puzzle 1

```
13  −      +  3   −
 −      +      +      +
    ×  16  ÷  14  −
 −      ÷      −      −
12  ÷  8   +  15  ÷
 +      −      +      +
    −      −      +  5
```
= 4

Puzzle 2

```
13  −  14  ÷      −
 −      +      +      −
    ÷  5   −  1   +
 ×      −      +      +
    +  16  ÷  8   −  9
 ÷      +      −      ÷
        −  11  +  3
```
= 5

Puzzle 3

```
 7  +      ÷      −  14
 +      −      ×      ÷
        −  3   ×  4   +
 −      +      +      ×
        −  2   ×  11  +
 +      −      −      −
 5  ×          −  15
```
= 6

In each 4 × 4 grid, place the numbers 1 [to 16] so that each row and column equals the target number.

Order of operations applies!

Sharpen your pencil, pop on your thinking cap and knuckle down!

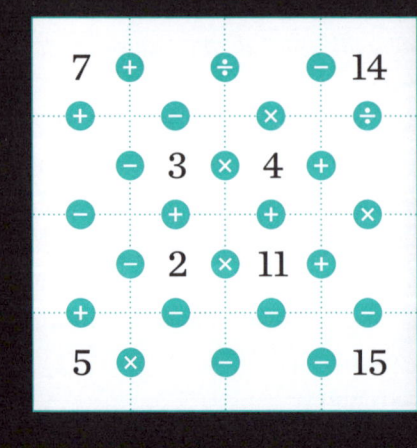

Farewell and good Knight

How's this for a book title: *What is the Name of This Book?*

That's the title of American mathematician and muso Raymond Smullyan's 1978 cracker which first introduced us to the 'Knights and Knaves' logic puzzle. You may recall in this puzzle some of the characters can only answer questions truthfully ... and others only falsely.

If you're a kid of the 80s, you might also recall this was a key scene in the 1986 fantasy classic (starring none other than David Bowie ...) *Labyrinth*. There, the protagonist finds herself faced with two doors guarded by creatures who are obviously Smullyan fans.

One door leads to the castle at the centre of the labyrinth and the other? Well, that door gets you the less desirable prize of certain death.

We're going to encounter our — er — Knavish and Knightish friends a few times here in *Numberland*, but let's kick things off with the dilemma faced by *Labyrinth*'s hero.

We are presented with the following: one guard always tells the truth, and the other always lies. We don't know which is which.[*]

Of course, we also don't know which guard is guarding which door (the one which leads to the castle, and the one which leads to [*buh, buh, buh, boom!*] certain death).

To figure out which door to choose, our hero may ask one guard one question.

So, how about it — who, and what, are you going to ask? You know where you'll find the answer ...

[*] If you want to see the original scene, you can find it on YouTube ... though why would you given my superlative description here?

Roujiamo

Like most Anglos, I'd always traced the history of the sandwich back to good old John Montagu, 4th Earl of Sandwich.

He was the 18th-century English toff who was said to like playing cards for so long into the night he would order his valet to bring him meat tucked between two pieces of bread. Over time his card playing buddies started to ask for 'the same thing as Sandwich!' They all got a bellyful, the cards didn't get greasy and the rest is history, as they say.

But it turns out the old sanger has been around for a lot longer than that.

The palm-sized 'Chinese hamburger' called 肉夹馍 — or *roujiamo*, meaning 'meat sandwich' — is a famous street food from the Shaanxi province. It's most commonly filled with minced pork that's been stewed for hours with over 20 herbs and spices (beat that, Colonel!). However, there are no set rules about how to make it, and alternatives abound, including a Muslim-friendly beef version from the Xi'an province, seasoned with cumin.

It's very tasty and, it turns out, very, very old. In fact, *roujiamo* lays convincing claim to being the world's oldest sanga since *baijimo* — the type of flatbread it uses — dates back to the Qin dynasty (221–206 BC) and the meat filling itself to the Zhou dynasty (1045–256 BC).

If you're curious to try one, pop down to your nearest Chinatown. You could do worse than heading to Cafe Mo'st in Sydney, where I first learned about this tasty old burger from a sign in their window. They also note that they've 'doubled the size'.

Roujiamo whopper, anyone?

The *Oxford English Dictionary* reports the first use of the word sandwich in this sense being from 1762 — so happy 257th birthday, sandwich.

Oh, and Cafe Mo'st, if you'd like to offer me free *roujiamo* for life as payback for this pretty solid plug, hit me up on Twitter @adambspencer.

Between this and the Lady Copa pizzeria on the Central Coast, my flatbread feasting for 2020 is *sorted*!

Going, going, gone ...

Here's a sobering statistic for you.

According to the United Nations, who recently completed its first global assessment of the natural world in 15 years, up to 1 million of Earth's estimated 8 million species face extinction — many within mere decades — as the Earth warms and humans spread further.

Hey, I don't want to get all preachy here but it certainly bears thinking about next time you're grumbling about taking out the recycling.

For more sobering stats on extinction in our own backyard, check out page 319.

If you're interested in reading the UN Report, you can check it out at www.un.org

ADAM SPENCER

Up to 20% of all power outages in the US are caused by ... squirrels!

If you've ever wondered where the biggest squirrel-related power outages in the US have been, you're in luck: the website CyberSquirrel1.com has you covered.

You won't be alone in your curiosity about these dastardly critters. John C. Inglis, former Deputy Director of the National Security Agency says, 'I don't think paralysis [of the electrical grid] is more likely by cyberattack than by natural disaster. And frankly the number one threat experienced to date by the US electrical grid is squirrels.'

As far as a public enemy number one goes … they sure are cute.

Sp-H-ooky

Let me tell you the tale of the ghostly 'H' ...

It's the 8th letter of the alphabet, and while there is one in the word 'aitch' and two in the word 'eighth', there is at least one where it shouldn't be ... in the word 'ghost'.

It appears that the H apparition first appeared when Flemish typesetters, who set up the printing press, added an 'h' since that's how the Flemish spelled it. For some reason, it caught on.

Not so for other Flemish words which include an extra H, such as ghoose, ghoat and ghirl.

And for any or any H-hounds out there wanting more H-appiness, there are 3 'aitches' in the word 'Shhh' (which is accepted by the international Scrabble dictionary) and 3 in highlight, hitchhike and shahtoosh (a light but warm Kashmiri shawl made with hair from the *chiru*, or Tibetan antelope, thanks for asking).

> In the unlikely event you're looking for a diabolically tongue-twisting limerick which plays heavily on the letter h, try this on for horrednousness:
>
> In Huron, a hewer,
> Hugh Hughes,
> Hewed yews of unusual hues.
> Hugh Hughes used blue yews
> To build sheds for new ewes;
> So his new ewes blue-hued ewe-sheds use.

'I love speculating about solutions to problems in mathematics. I have no interest whatever in sudoku. But I do look at chess and bridge problems in newspapers. I find that relaxing.'

—Vikram Seth

One of the world's great novelists and a nerd after my own heart

ADAM SPENCER

The Quiet Americans

The year is 2016.

US diplomats in Cuba are under siege from a dastardly sonic weapon developed by — who knows, maybe the Russians? — causing headaches, nausea and even hearing loss. Further examinations show signs of concussion and other brain injuries. The threat must be neutralised. But how?

It sure sounds like the précis of a James Bond novel, but it turns out the truth is far more Austin Powers.

After high-level protests from the US Government and amid international uproar, researchers Alexander Stubbs of the University of California, Berkeley, and Fernando Montealegre-Z of the University of Lincoln in England, studied a recording of the cruel sonic device made by diplomats and published by the Associated Press.

'There's plenty of debate in the medical community over what, if any, physical damage there is to these individuals,' said Stubbs in a phone interview with *The Independent* newspaper. 'All I can say fairly definitively is that the AP released recording is of a cricket, and we think we know what species it is.'

That would be the Indies short-tailed cricket, whose 'pulse repetition rate, power 27 spectrum, pulse rate stability and oscillations per pulse' were a pretty good match.

'An echoing cricket call,' they noted in their report, 'rather than a sonic 33 attack of other artificial device, is responsible for the sound released in the recording.'

No word on whether we're training cicadas to infiltrate enemy positions at the moment.

Worth a shot?

So in this book we've outed squirrels as the enemy of the US power grid and crickets terrorising international diplomacy.

Any chance we could get the squirrels and crickets to just sit down with us at the negotiation table and all be friends again?

YOU ARE HERE

Knight time in *Numberland*

Journeying through *Numberland*, we once again find our way blocked.

Recall, if you will, our riddling knight and knave back at page 159. We know that *Numberland* is populated by two kinds of people: knights who always tell the truth, and knaves who always lie.

This fine and sunny day (it is where I am at the moment, sorry to hear it if it's cold and gloomy where you are) we come across a pair of cards. The first says the other card is a knave. The second wrinkles his nose and says, 'Neither of us are knaves!'

Tell me, what are they *really*?

Head to the back of the book to check your answer.

HINT: Remember our crash-course in Boolean algebra? If you take nothing else from it, try using labels for the knights and knaves we come across: A and B. So here, A says that B is a knave; B says that neither A nor B are knaves.

ADAM SPENCER

Octopus ♥³ you

We all know that octopuses have 8 'arms'.

But did you know that they have 3 hearts? One is called a 'systemic' heart and circulates blood around the octopus's body. This heart is helped out by two other 'branchial' hearts that pump the blood through each of the octopus's gills.

Octopus blood is blue — this is not because they were born into the upper class and went to the right sort of school — but because their blood contains copper-rich, blue haemocyanin. Our blood contains iron-rich haemoglobin in red blood cells.

Oh, by the way — octopuses have 9 brains. One in their head to control their nervous system, and one in the base of each of their 8 arms.

The giant Pacific octopus is probably the world's largest. One individual specimen was recorded weighing 71 kilograms.

More score

| 7 | | 8 | | 8 | | 6 | | 5 | = 2

| 8 | | 5 | | 7 | | 6 | | 4 | = 2

Reach the goal number by creating an equation using the provided numbers and your own choice of operations.

Numbers are placed in white squares, while operations (+, −, ×, ÷) are placed in green squares.

Order of operations matters here, per usual, and you can't use brackets!

Read the numbers off in order for your final score.

| 9 | | 6 | | 6 | | 7 | | 4 | = 2

Again, I've given you the equation that gives the highest score. But this time, you have to work out the operators yourself.

Tank-tastic

In previous books, I've shared the famous equation:

$$\frac{(12 + 144 + 20 + 3\sqrt{4})}{7} + 5 \times 11 = 9^2 + 0$$

Now you can check that:

$$\frac{(12 + 144 + 20 + 3 \times 2)}{7 + 5 \times 11} = \frac{182}{7 + 55} = 26 + 55 = 81 = 9^2 + 0$$

So yes, the equation holds. But surely that alone is not enough to make this special?

The fame derives from the fact that you can read this equation as a poem, more specifically a limerick. Using some old fashioned terminology for 144 (a gross) and 20 (a score) we get:

A dozen, a gross, and a score,
plus 3 times the square root of 4,
divided by 7,
plus 5 times eleven,
is nine squared and not a bit more.

* For more on the beauty of 'mathematical poetry', check out 'Pi-kus' on page 93.

It turns out this isn't the only example of mathematical poetry floating around out there.

In 2017, the *Journal of Humanistic Mathematics*, part of a project hoping to bridge mathematics and the arts, called for mathematical *haiku* ... 3-line poems in the '5-7-5' syllabic form that expressed a mathematical idea or experience, and hopefully connected it to the human condition. In deference to traditional Japanese *haiku*, the journal encouraged poets to consider using allusions to nature or the seasons in their work.*

They were inundated with examples. Here are a few corkers.

Valentina Ranaldi-Adams
square the radius
and then multiply by pi —
full moon in autumn

Blaine Schmidt
A squared plus B squared
Results in C squared each time ...
Pythagorean

Pi by Benjamin van Duin
Infinite digits
Pi is never repeating —
And inedible

And a slightly racier offering by Greg Warrington
Fibonacci's law
feeds sequential offspring from
two horny rabbits

Even one entitled 'Index Theory' thanks to Francesca Arici[*]
a Dirac operator
recovers the manifold's topology
via its fredholm index

Which you probably don't need me to tell you is a little bit too heavy to delve into here.

Up until this point I'd heard about the traditional 5-7-5 syllabic form poems. But it wasn't until I spent a week in Tokyo recently that I became aware of the breadth of poetic forms in Japanese culture. There's even a museum dedicated to poems written by the Emperor Hirohito and his wife Michiko that reveal their feelings about certain events they witnessed.

It evidences a beautiful aspect of Japanese culture. By comparison, I could never imagine Donald Trump capturing the solemnity of a civic occasion in a poem. Maybe a tweet with LOTS OF CAPITALS, but never a poem.

So, without further ado, here is my beginner's guide to Japanese poetic forms. I love the words, but the nerd in me also loves the numerical underpinnings.

[*] Look, this one is actually in 7-11-7 form, but how can I pass up such a nerdy *haiku*?

Mora less ...

In Japanese, the *mora* (pl. *morae* or *moras*) is a rhythmic unit known as *on* (音) or *haku* (拍).

The Japanese writing system of *kana* is based on *morae*, placing one *kana* on each *mora*. Although *morae* are subtly different from syllables, anglophones have generally interpreted these units of speech sounds as syllables — leading to the common (mis)understanding of Japanese poetic forms as being based on 'syllable counts'. The *haiku*, for example, is not really a 17-syllable poem, but a poem with 17 *morae*. Because the difference is subtle, I will use *morae*/*on*/syllables interchangeably in this chapter.

Here are 6 Japanese poetic forms.

Haiku

A *haiku* is a 3-line poem, traditionally made up of 17 syllables distributed across 3 lines in a 5-7-5 *moraic* structure. *Haiku* are characterised by references to nature and the seasons and often involve *kiru*, a juxtaposition of two images or ideas with a *kireji* (a cutting word) between them.

More recently, the modern *haiku* does not strictly follow 17 syllables in 5-7-5 form. Some *haiku* poets follow a 5-3-5 form, whereas some do not even follow the uniform pattern of syllables ... those crazy poets.

Here are two classic *haiku* poems from Basho Matsuo (1644–1694), who, as we heard earlier, is said to be the greatest *haiku* poet:

An old silent pond ...
A frog jumps into the pond,
splash! Silence again.

Autumn moonlight —
a worm digs silently
into the chestnut.

Great stuff, Basho.

Senryu

Senryu is what most people mean when they refer to *haiku*. Like the *haiku*, the *senryu* form features 17 *morae* in a 5-7-5 form. The basic difference is that *haiku* focus on seasons and nature, whereas the *senryu* is about the ironies of life. Thematic treatment in *haiku* is serious whereas *senryu* are humorous or cynical.

How about some examples? Glad you asked.

UNTITLED
Ecstatic at being
set free,
the bird collides with a tree.
— JC Brown

Ureshisa no
ki ni tsukiataru
hanashi-dori

A CRUEL JOKE
Nursing a fracture
On bed, leg stretched out in cast;
Friends smile and sign ...
— Mamta Agarwal

DONALD TRUMP CUTOUT
climate march
the pink-haired girl face to face
with a Trump cutout
— Chenou Liu

TRUMP TIMES TWO
father a builder
his daughter, true diplomat;
beauty and the beast
— Govind Ramakrishnan

I believe that's one the kids call a 'sick burn' ...

Dodoitsu

Dodoitsu is a form of Japanese 'folk verse' usually written about work or love and often with a touch of humour. *Dodoitsu* poems consist of 4 lines with the *moraic* structure 7-7-7-5 and no rhyme for a total of 26 *morae*, making it one of the longer Japanese forms.

For example:

Thursday afternoon is dead,
phones become silent and wait
until minutes before 5
on Friday to ring
— Judi Van Gorder

Tanka

Tanka or 'short song' poems are composed of 31 syllables in 5-7-5-7-7 form.

Take it away, Can Sonmez:

Subtle hints of spring
In the wet bark of the tree
Dew dripping from leaves
Then runs down the russet trunk
Pools round the roots and is drunk

Katauta

The *katauta* is a 3-line poem which, in its common form, has 19 syllables in 5-7-7 form. *Katauta* poems are specifically addressed to a lover.

When paired together, multiple *katautas* act as a question-and-answer conversation between lovers to form the 38-syllable *sedōka* poetic form.

So Robert Lee Brewer wrote this 'Untitled' *katuata*:

why do winter stars
shine brighter than summer stars
as if they are shards of glass?

... which he expanded to form an 'Untitled' *sedōka*:

why do winter stars
shine brighter than summer stars
as if they are shards of glass?

don't blame the seasons
on the ever changing heat
of your lover's quick embrace.
— Robert Lee Brewer

You old smoothie, RLB!

Gogyohka

A comparatively recent invention, proposed by the poet Enta Kusakabe in 1957, the *gogyohka* (literally, '5-line poem') is an offshoot of the *tanka* form with very simple rules: the poem is comprised of 5 lines with one phrase per line.

There are no syllabic or *moraic* constraints; the length of each line, as well as the poem's tone and subject focus, is left to the writer's discretion.

Three people
listen to the doctor's explanation.
He said this, he said that
but
what did he actually say?
— Haru

And I found this rather fruity example online:

My younger sister
says
'I like everything about Nao'
Does she like his 'intestines', 'bowels'
and 'anus'?

... which was allegedly written by Onishi Maho, a second year primary school student.

I think we might leave it there ...

ADAM SPENCER

To Express *e*, Remember To Memorise A Sentence To Memorise This

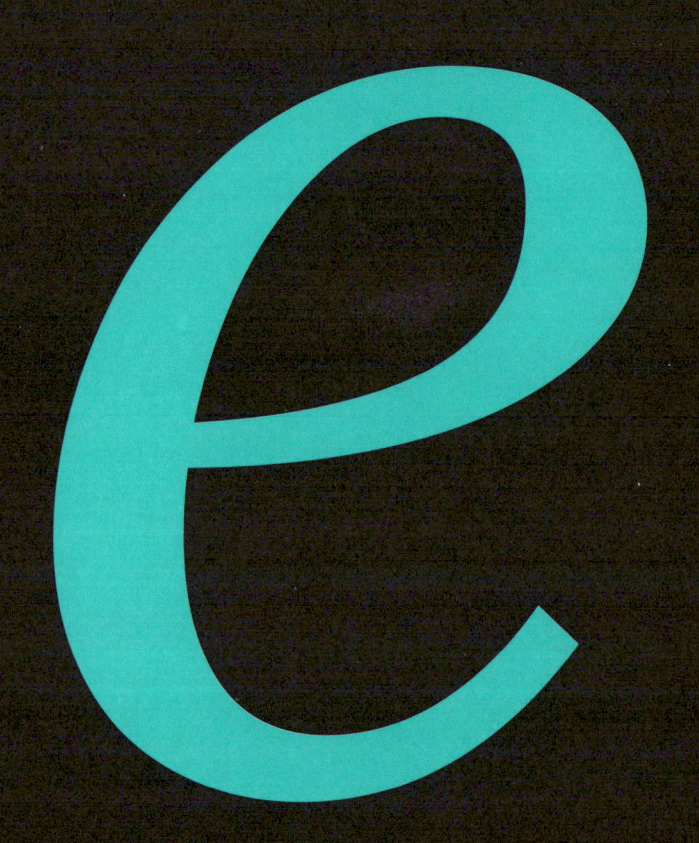

e

YOU ARE HERE
178

If you've ever read any of my books, heard me speak, seen my nerdy T-shirt collection or met me ... you'll probably know that I'm a bit of a #fanboy of the greatest EVER mathematician, 18th century Swiss genius, Leonhard Euler.

The seemingly nonsensical mnemonic on the opposite page is a handy way to remember Euler's constant (e) to 10 decimal places: the number of characters in each word gives us the digits.

e = 2.7182818284...

Pony express

When you're 7 feet tall and weigh somewhere in the order of half a tonne, you don't want to be flying economy.

That's why some of the world's most valuable racehorses choose to fly in style, air-conditioned to a horse-friendly 13° Celsius, on a Boeing 727-20 whose livery reads 'First Class Equine Air Travel'.

The plane is — not surprisingly — highly customised to accommodate the choosiest nag and carries, at any given time, a cargo worth in excess of US$30 million.

But I've got to be honest with you, my favourite fact about this flying punny pony express is its name: *Air Horse One*.*

*If you don't get the hilarious pun, *Air Force One* is the name of the plane that ferries around the American President ... *and* of a 1997 political action movie thriller starring Harrison Ford as POTUS.

Graphene

You might recall we've visited the amazing properties of graphene more than once over the years in my books.

If not, go and grab them (they're greaaaat!) but here's a quick primer: graphene is the strongest, lightest material discovered. It's essentially a sheet of carbon atoms arranged in a very special pattern which you can pick up, and yet is quite transparent and 'stretchy' (beyond about 20% of its original length). It beats diamonds in thermal conductivity and is so impermeable that even helium atoms can't squeeze through it.

It's very impressive stuff.

But far from simply being another expensive way to make spoilers for expensive cars driven by highly paid footy players, researchers at Manchester University have invented a sieve made from the wonderstuff which can turn saltwater into drinking water.

Given that 1 in 10 people on Earth don't have access to safe drinking water, that's an even more impressive feat.

Snake tales

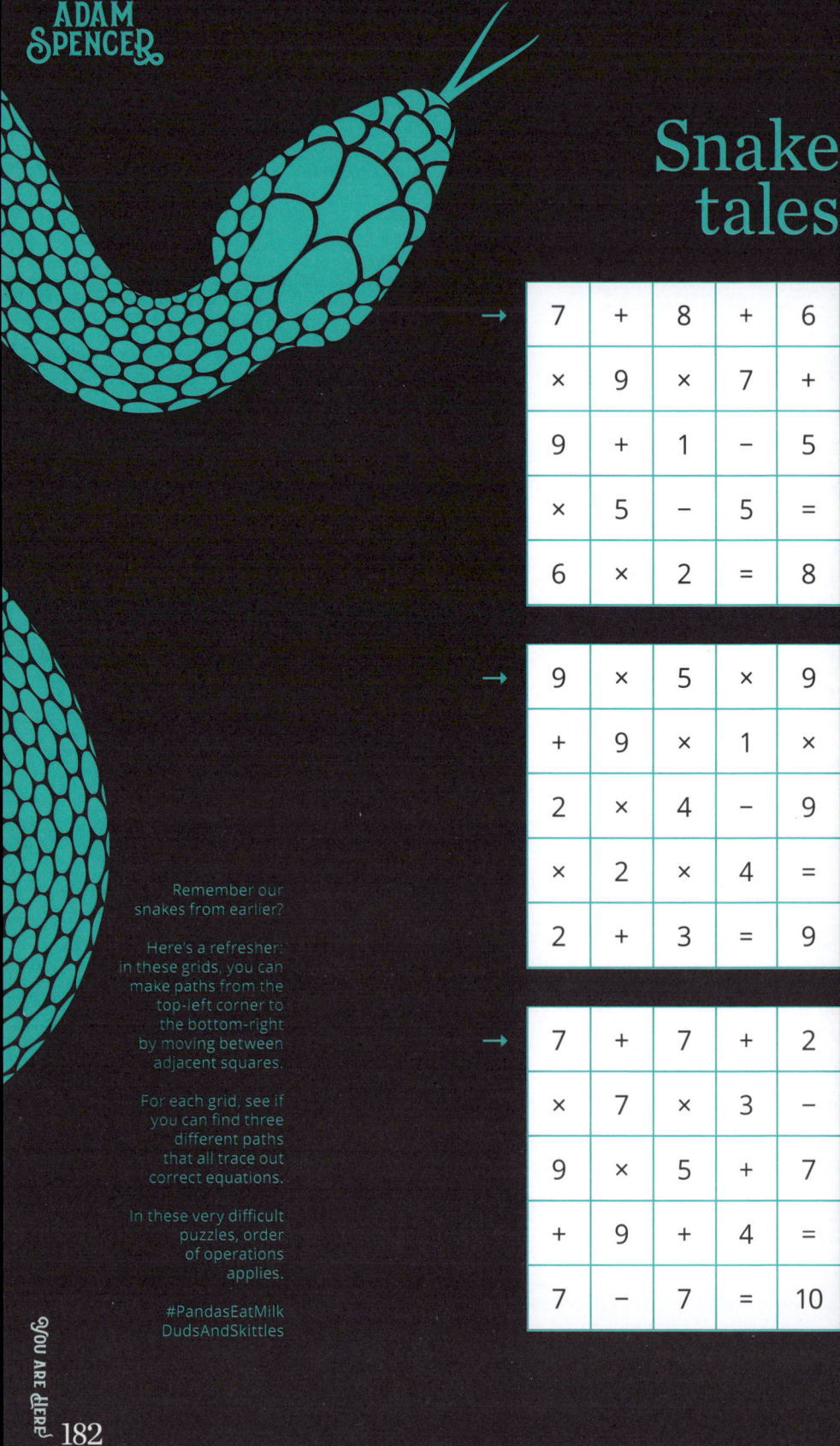

7	+	8	+	6
×	9	×	7	+
9	+	1	−	5
×	5	−	5	=
6	×	2	=	8

9	×	5	×	9
+	9	×	1	×
2	×	4	−	9
×	2	×	4	=
2	+	3	=	9

7	+	7	+	2
×	7	×	3	−
9	×	5	+	7
+	9	+	4	=
7	−	7	=	10

Remember our snakes from earlier?

Here's a refresher: in these grids, you can make paths from the top-left corner to the bottom-right by moving between adjacent squares.

For each grid, see if you can find three different paths that all trace out correct equations.

In these very difficult puzzles, order of operations applies.

#PandasEatMilkDudsAndSkittles

NUMBERLAND

'If you stop at general math, you're only going to make general math money.'

—Snoop Dogg

Fo' shizzle

Do you want to change your password?

R umour has it ...

That for almost 20 years the US nuclear launch code was 00000000. You can read more about that in my awesome, available-where-all-good-books-are-sold or on-my-website bestselling book *The Number Games*.

But before you snort and roll your eyes ... hands up if any of the following look, ah, somewhat familiar to you?

123456
123456789
qwerty
password
111111
12345678
abc123
1234567
password1
12345

Consider 20 character passwords that are case sensitive, (so b is different to B) and can include letters, numbers and the 10 symbols above the numbers on a standard keyboard.

You have 26 + 26 + 10 + 10 = 72 possible characters for each of the 20 positions. This creates 72^{20} or roughly 14,000,000,000,000,000, 000,000,000,000,000,000 possible passwords.

Though one of those would have to be ADAMSpEnCeRisAWESOME (which is my password, so back off!)

Still feeling smug? Well, if 'monkey', 'dragon' or 'iloveyou' or any number of combinations of swear words ring bells, you probably need to change your password, buddy.

******|

The UK's National Cyber Security Centre devised the above top 10 by analysing the passwords belonging to breached accounts worldwide. It will astonish you, I'm sure, to learn that there were no 20-string random letter and number combinations there.

Look, I *am* judging you a little if your password of choice is 123456. But at least you're in good company: 23.3 million hacked accounts studied also used that high security top secret code.

Around 7.7 million people decided to protect their embarrassables with 'qwerty', while 3 million people felt 'password' would be sufficiently difficult for hackers to unravel.

The most common names were Michael and Ashley, and — sigh — Blink182 was the most popular musical artist. Liverpool topped the Premier League of passwords and cowboys1 — presumably in honour of the NFL's Dallas Cowboys — ranked highly in the US.

Two factor authentication, anyone?

Lots of pi

The history of calculating pi is littered with hardcore heroes.

Ever since Indian mathematician and astronomer Madhava of Sangamagrama discovered the infinite series for π, now known as the Madhava-Leibniz series, geeks around the world have raced to calculate pi to ever increasing decimal places.

In March 2019, Japanese computer scientist (and Google employee) Emma Haruka Iwao made a pretty impressive entry into the race by calculating pi to 31,415,926,535,897 — that's just over 31.4 trillion! — digits. In doing so, she crushed the nearest calculation from 2016 by trillions of digits. No biggie.

Far from using bespoke programs on supercomputers, Iwao used an 'off the shelf' multi-threaded program 'y-cruncher', the home pi calculator's weapon of choice, across 25 Google Cloud virtual machines ... which by no small coincidence is her area of expertise.

It took 121 days and generated 170 terabytes of data — roughly the equivalent to the amount of data in the entire US Library of Congress print collection — and was celebrated by Iwao and her colleagues in her Tokyo office ... with actual pie.

Well deserved!

You've probably noticed she stopped exactly at the point where the number of digits matches pi itself to 13 decimal places.

If you want to give the record a whirl, you too can download a y-cruncher. Point your browser to http://www.numberworld.org/y-cruncher/#Download and break a leg!

Fame, glory ... and a bit of history

T he top gong in maths for 2019, the Norwegian Academy of Science and Letter's Abel Prize ...

... went to Karen Uhlenbeck, an emeritus professor at the University of Texas, kickass numbers geek and ... woman.

About time, too. This is the first time in the award's 16-year history that it's gone to a number cruncher of the fairer sex.

She received it for 'the fundamental impact of her work on analysis, geometry and mathematical physics' which included research into the complex shapes of soap films.

Not just your average sudsy dishwater, mind you; we're talking abstract, high-dimensional curved spaces, which was a precursor to her work describing fundamental interactions between particles and forces.

Dr Uhlenbeck, 76, gets fame and glory as her reward.
Oh, and a cool million bucks (Australian).
Played, U-beck.*

*Another nickname I came up with, and which I doubt very much the good doctor chooses to use herself.

The legs of a Japanese spider crab can grow up to 12 feet long

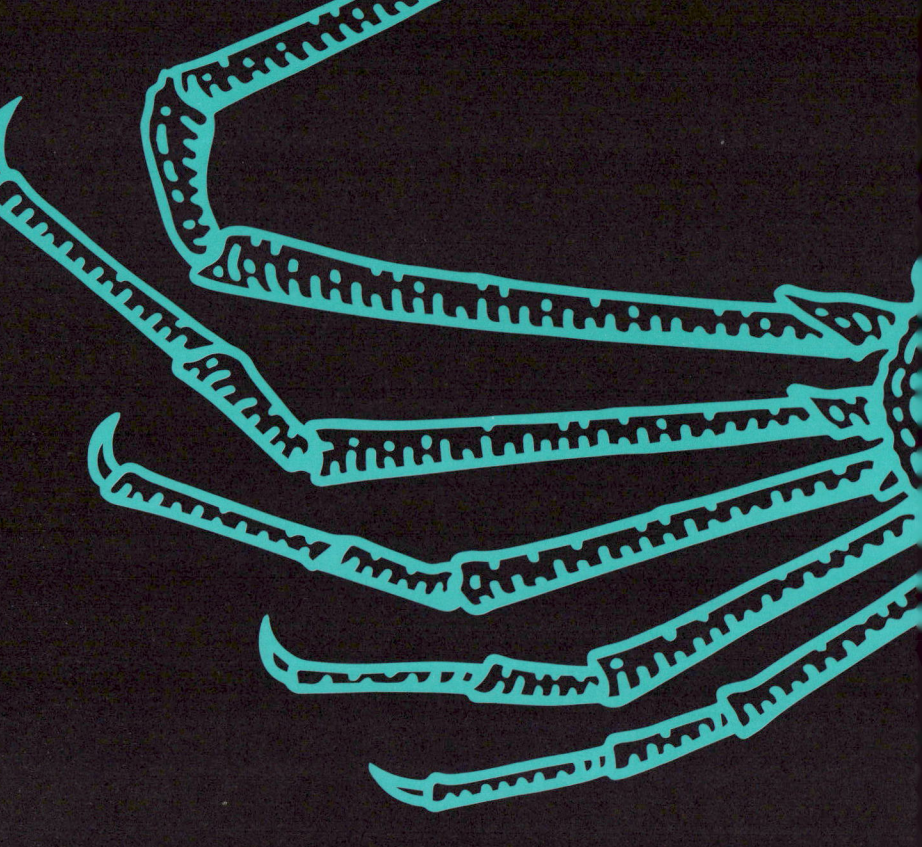

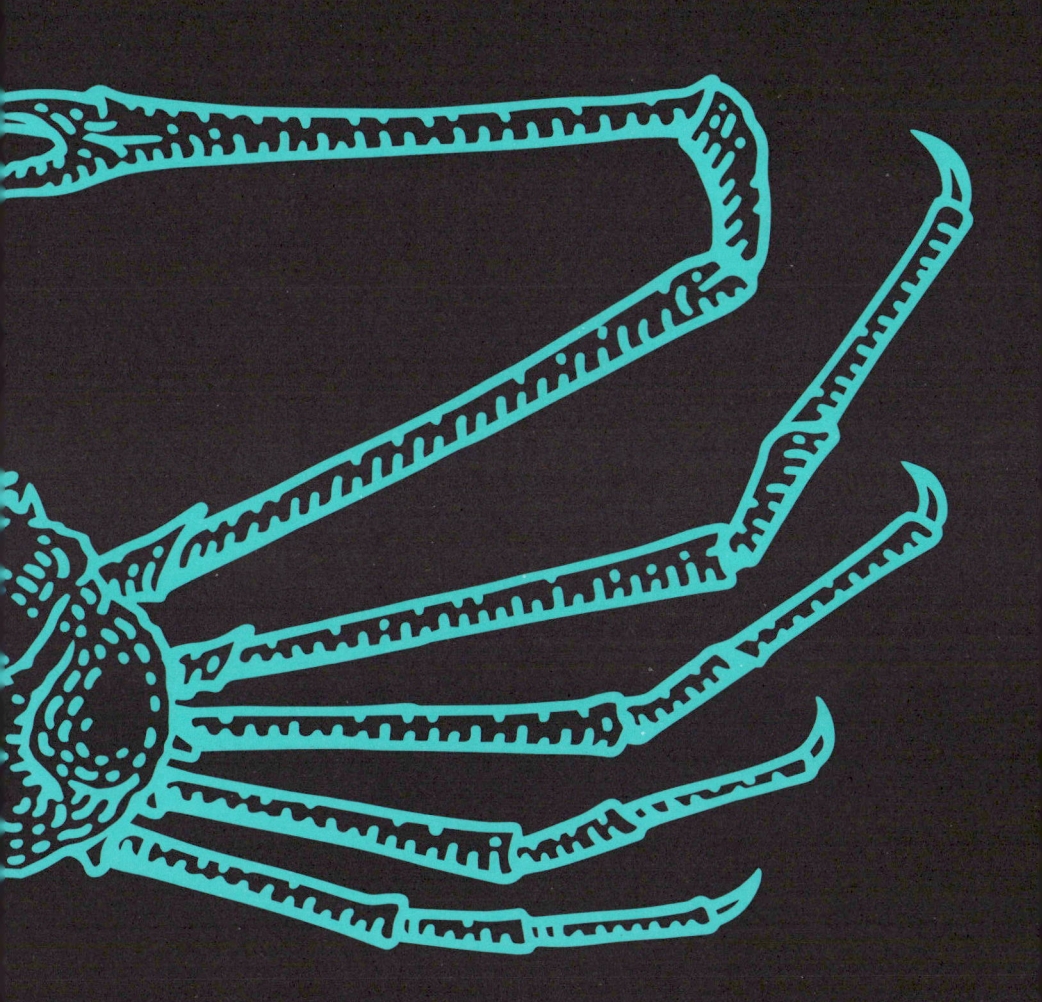

That's a little over 3.5 metres,
for the metrically minded ...

The great Auto-Tune debate

If you're a music lover, you've almost certainly listened to an 'auto-tuned' vocal without even knowing it.

But I bet you never knew about the controversy surrounding the tech ... and the genius behind its invention.

Back in the 1970s, an American kid named Andy Hildebrand was noodling around, disrupting class, and trying to improve his grades. After a few false starts, he turned things around, did better, and got into science. Big time.

In 1976, he was awarded his electrical engineering PhD, excelling in the areas of linear estimation theory and signal processing. Soon enough, he landed a lucrative job with Exxon — yep, *that* oil company. But he didn't stay there long. In 1979, he left Exxon and founded a company called Landmark Graphics which pioneered a way of mapping the Earth's subsurface by processing thousands of lines of data, creating cool 3D seismic maps.

But what's this got to do with music, Adam? Or maths, for that matter?

Well, after Hildebrand sold Landmark Graphics for around US$525 million, he decided to return to his very first love, music. As a kid, he was brilliant on the flute, and so once he had the money in the bank, he began studying composition and mucking around with synthesisers.

He soon ran into problems: when he tried to make his own samples (of his flute), the quality of the sound was awful. The machines he was using simply couldn't hold enough data. So he created a processing algorithm that condensed the audio data, giving a much nicer and smoother sound. He packaged his algorithm into a software program called Infinity and started giving it to composers.

Let's just say the effects of Infinity were ... seismic! The program improved digitised orchestral sounds so much it upended Hollywood's music production world. Composers could now accurately reproduce sounds digitally — without the need for expensive musicians.

But because Hildebrand gave it away, competitors quickly copied it, and he didn't get to add much to his US$525 million piggy bank.

But wait, there's more. In 1995, scouting for something new to invent, he asked a bunch of musicians what needed to be created. Someone piped up and said, 'Why don't you make me a box that will let me sing in tune?'

Hildebrand got to work using similar algorithms to that which he'd used for Infinity. The programming incorporated some pretty heavy-duty maths: correlation (statics determination); linear predictive coding (deconvolution); synthesis (forward modelling); formant analysis (spectral enhancement) and processing integrity to minimise artifacts.

While working for Exxon, when dealing with massive datasets, Hildebrand had employed autocorrelation (an attribute of signal processing) to examine not just key variables, but all of the data, to get much more reliable estimates. He soon realised this could apply to music too:

'When you're processing pitch, you add wave cycles to go sharp, and subtract them when you go flat. With autocorrelation, you have a clearly identifiable event that tells you what the period of repetition for repeated peak values is. It's never fooled by the changing waveform. It's very elegant.'

The result was the Auto-Tune program which debuted in 1996 and meant that anyone could sing in tune. Instead of a singer doing hundreds of takes to get the song right, Auto-Tune came in and subtly and unobtrusively corrected notes that were just a bit off-key. What might have cost a recording engineer tens of thousands of dollars to manually correct, could be done in an instant.

Well, the backlash was inevitable. Musicians far and wide said the tech was fake and that 'real' music shouldn't need such gimmickry. Hildebrand says in response that all electronic recording is essentially 'fake' — if you're going to complain about Auto-Tune, you might as well complain about speakers and synthesisers and recording studios themselves. The debate rolls on. But perhaps the last word should come from the genius inventor himself: 'Sometimes, I tell people, "I just built a car, I didn't drive it down the wrong side of the freeway!"'

'Wherever there is number, there is beauty.'

—Proclus

Take this with a grain of (pink) salt ...

Everyone's talking about Himalayan pink salt, right?

I mean, I can't go 10 metres outside my front door without someone bailing me up in the street and dumping the dirt on this miracle substance.

And don't get me started on the 70,000+ #pinksalt images on Insta.

Seriously, the pink salt fad may not have reached every city around the world, but a lot of claims are being made about the health benefits, so I reckon we need to do a bit of myth-busting.

First up, 'Himalayan pink salt' doesn't actually come from the Himalayas. Most of it is mined at a huge salt mine in Pakistan ... hundreds of kilometres west of the actual Himalayan mountain range. It's not a new mine either — some sources claim people have been gathering pink salt there for more than 2000 years.

But what about the health benefits, Adam? And the pretty colour!

Pink salt gets its cool colour from the minerals it contains, including magnesium, potassium and calcium. But it doesn't have a monopoly on those trace minerals and they're available in plenty of other places — like fruits and veggies. Okay, but pink salt producers (and wellness gurus) tell us that pink salt can provide the body with all 84 natural elements, benefitting our pH levels, hydration, weight, hormone balance and sleep patterns ...

Hmm, maybe.

The catch is you'd have to eat an enormous amount of the stuff to get any measurable benefits. And pink salt can cost 30 times more than regular white table salt.

So from a dietary point of view, pink salt is no better or worse than the white stuff. Good old fruits and vegetables offer all the same minerals — and are much better for you — and of course health professionals are urging us to consume less sodium over all, not more.

So, there's the drum on the pink stuff.

If I were you, I'd take Himalayan pink salt ... with a grain of salt.

And in case you needed any other reason to ease up on the white powdery stuff, a report released in June 2019 by South Korean researchers and Greenpeace Asia found that 90% of tested brands of table salt ... contained microscopic plastic particles.

Jeopardy!

Few people have dominated the US gameshow *Jeopardy!** quite like James Holzhauer.

In April 2019, the 34-year-old, pro sports gambler from Las Vegas (where else?) wielded the buzzer deftly enough to amass only the second million dollar haul in the show's history. Nothing to be sneezed at for a show which debuted on 30 March 1964. So is being a professional gambler an advantage in what is otherwise a trivia game? Turns out it might well be.

Holzhauer utilised a strategy to go for the high-value clues first, look for bonus 'Daily Doubles' (hidden behind one clue in *Jeopardy!*, and two in *Double Jeopardy!*) and, when finding them, go 'all in' with his bet — effectively enabling him to double his already huge lead. Old-school Jeopardists start low and work up to higher value questions. J-Hol flipped that strategy on its head. And it's worked pretty well for him.

Holzhauer still holds the record for single-game winnings: $131,127 on 17 April 2019. He added his name to the list again on 23 April ($118,816). And, ordered by prize, he also joined the top 10 list on 9 and 16 April. Oh, and 1 May, 30, 25, 22, 12 April, 25 April plus 3 May. At the time of writing, he was ranked number two in overall regular-play winnings and games consecutively won, behind legend Ken Jennings (74 games in 2004, total winnings $2,520,700). He was finally defeated on 3 June by 27-year-old contestant Emma Boettcher. His winning streak was 32 games**.

So his gambler's instinct helps. But what about the trivia part? Holzhauer told ESPN that reading children's books is a more effective strategy than textbooks since they cater to readers who are unlikely to be too naturally interested in the subject.

Oh, and the $131,127 question? *'His first name refers to the ancient district in which you'd find the Greek capital; his surname is a bird.'* Answer's at the end, of course.

---

* If you haven't seen it, the show works something like this. Three contestants, including the returning champ from the last round, compete in 3 quiz rounds: *Jeopardy!*, *Double Jeopardy!* and *Final Jeopardy!* Each round contains 6 categories of 5 clues, covering a broad spectrum of interests — from the arts to pop culture and a lot in between. And of those clues, each is ostensibly valued in dollar terms by its supposed difficulty. Got it?

** Over his 33 appearances, the Holz answered an astonishing 97% questions correctly!

Operation!

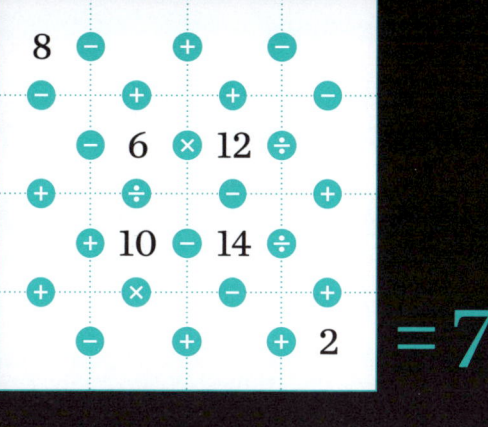

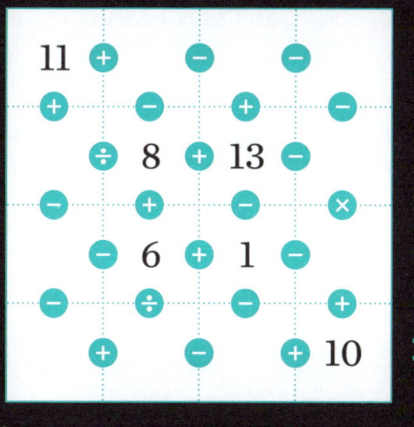

= 7

= 8

= 9

For each 4 × 4 grid, enter the numbers 1 to 16 so that each row and column equals the target number.

Order of operations applies!

These are the fewest hints you'll be getting, so sharpen your pencil, strap your thinking cap on and knuckle down!

Genus *Giraffa*

Our biggest ruminants and tallest animals on land have a pretty fascinating story.

Giraffes are obviously some of the most magnificent creatures around — if you go in for long necks and lots of spots, that is. And they're also rare: they're categorised as 'vulnerable to extinction' with only around 97,000 animals left in the wild.

But what particularly interests me is the fact that the International Union for Conservation of Nature recognises just one species, *Giraffa camelopardalis*, with 9 subspecies. Why is this interesting? Well, because for such a distinctive looking animal that's been studied for decades, there's still a lot of conjecture as to just how many species exist.

Back in 2001, scientists suggested there were two species. Then in 2007, others suggested there were in fact 6 species. In 2011, this increased to 8 species. And in 2016, scientists said there were in fact multiple species. The researchers suggested 4 of those species had not swapped genetic info for 1 to 2 million years! And by 'swapped' I think you know what I really mean.

So while science has advanced our knowledge of the natural world in myriad ways, we still have so much more to learn. And while the jury is still most definitely out on the exact number of living species of giraffe, there's no doubting there are at least 7 *extinct* species. Which is a sobering thought.

A group of giraffes is called a tower. When the males are fighting for a mate they go the knuckle by swinging their necks and headbutting each other ferociously. Humans have been killed by these savage blows.

Extraterrestrial cyclones

If that chapter heading hasn't grabbed your attention, nothing will. Except maybe for the *next* riveting chapter heading in this book!

You may be forgiven for thinking that cyclones and other similarly 'Earth-like' weather events only occur on our blue planet. Well, I'm here to tell you that others in our solar system get to experience windy days too. Yay.

Venus has two huge pairs of 'anticyclones' swirling at the planet's poles. The southern vortex is about the same size as Europe and does its business between about 40 and 60 kilometres above the planet's surface. Sometimes the vortices split into 2 and 3 smaller vortices which swirl around each other forming really cool patterns.

But I reckon Saturn has some of the best polar vortices. Back in 2009, the *Cassini-Huygens* spacecraft took photos of an amazing hexagonal cyclone at the planet's north pole measuring some 25,000 kilometres in diameter and with a depth of around 100 kilometres.

Just another reason to love the gas giant.

Garbology

Juan Pujol García was one heck of a character.

Born in Barcelona in Spain in 1912, he got caught up in the Spanish Civil War which gave him a pretty strong dislike of Communism and fascism.

In 1939, as World War II engulfed Europe, he decided he had to do something for his fellow citizens. In such situations, you and I might donate extra clothes or foodstuffs to the war effort. Maybe offer to billet a refugee or two.

Juan had bigger ideas. He approached the British and US intelligence services and asked if he could become a spy for them! When they said 'no' 3 times, he decided he'd become a double agent. He approached the Germans, who said, '*Ja bitte!*'

They told him to go to Britain and dig the dirt. Instead, he went to Lisbon in Portugal and created a whole load of fictitious material to put the Germans off the scent.

Juan went on to invent a nifty network of fictitious sub-agents who could be blamed for all sorts of false info. When the Germans spent a lot of time and resources tracking down a make-believe convoy as a result of Juan's work, the Brits and other Allies finally took notice and accepted him into the official spying fold, giving him the code name 'Garbo'. Bit harsh.

Juan and another double agent Tomás Harris spent the remainder of the War in Britain expanding their make-believe network of spies. In the end, the Germans were paying for no fewer than 27 non-existent agents.

Juan even played a role in 'Operation Fortitude', which misled the Germans about the timing and location of the Normandy landings in 1944. For his significant efforts, he was awarded the Iron Cross and became a Member of the Order of the British Empire.

Eat your heart out, 007.

Tile away, tile away, tile away

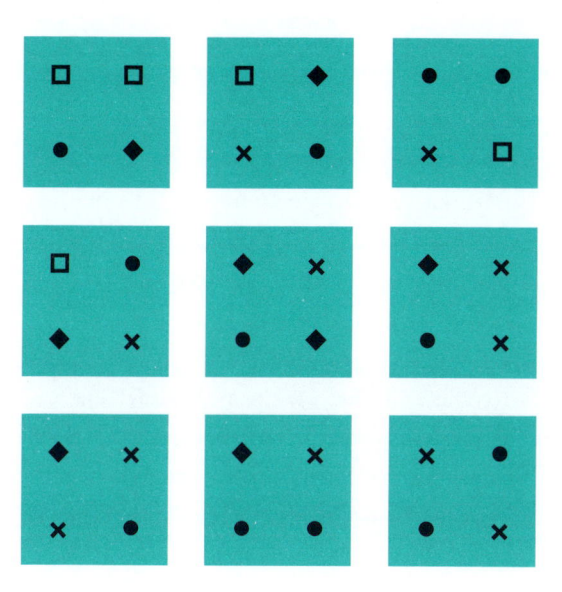

Think you're done tiling the bathroom? Think again, my friend!

Remember to place the 9 tiles into a 3 × 3 grid so that adjacent tiles show the same symbol along their touching edges.

Now we go up a gear. The tiles may need to be rotated (though never reflected) to find the solution.

In both instances, the top right tile is in the correct position and facing the right way.

Grab the grout and tile on!

As always, you can head over to my website to download a cut-uppable copy of this puzzle.

Point your browser to adamspencer.com.au

201

I'll be seeing you, *Opportunity*

All good things must come to an end. That's the old saying, at least. But no one expected NASA's incredible Mars rover *Opportunity* to survive quite as long as it did.

The 6-wheeled vehicle landed on the red planet back in 2004 with a mission to spend around 3 months there gathering information. It followed in the footsteps (wheel-tracks?) of the earlier NASA rovers, *Curiosity* and *Spirit*. Sadly both of those vehicles bit the dust. Literally!

But *Opportunity* was the little engine that could, clocking up more than 45 kilometres around Mars over a 15-year period (or 5111 Martian days which are about 40 minutes longer than an Earth day). It helped discover that water once existed on Mars, which is pretty massive.

But sadly, in January 2019, NASA announced that *Opportunity* had stopped responding to their flight controllers. The controllers tried again and again to make contact and even transmitted a recording of the late, great Billie Holiday's track 'I'll Be Seeing You' in one last valiant attempt to wake it up. But to no avail. *Opportunity* had been taken out by one of Mars's ferocious sand storms.

Hey, I don't want to leave you with a sad ending. I'm thrilled to say that in November 2018, the 3-legged NASA spacecraft *InSight* successfully landed on Mars following a 6-month, 480 million-kilometre trip. Good luck with the mission, *InSight*.

Puzzle pestilence

In 1880, the 'Fifteen Puzzle' was sweeping the United States.

A bit like Fortnite, but without electricity, high-powered rifles or exploding heads, this little sliding puzzle consisted of a frame of numbered square tiles in random order with one tile missing. The aim o' the game was to place the tiles in order by making sliding moves that used the empty space.

Sounds fun, right? Well, no. According to the *New York Times*, who I think were taking the mickey a little bit:

'No pestilence has ever visited this or any other country which has spread with the awful celerity of what is popularly called the "Fifteen Puzzle" ... it now threatens our free institutions, inasmuch as from every town and hamlet there is coming up a cry for a "strong man" who will stamp out this terrible puzzle at any cost of Constitution or freedom.'

In case you're wondering, 'celerity' means movement at great speed — don't worry, I had to look it up too.

I think someone at *The Times* needed a good lie down.

Three mathematical facts about the Fifteen Puzzle:

1. If you swap the 1 and 2 in a correct Fifteen Puzzle, you cannot solve the new puzzle. In fact, take a solved puzzle, move the pieces for a while, then swap any two tiles and you have a puzzle that cannot be solved.
2. If you take all the tiles out and reset them randomly, exactly 50% of the time your new puzzle will be solvable, the other 50% of the time you can get it to read 2 1 3 4 5 6 ... but not solve it.
3. The number of possible positions of the Fifteen Puzzle is $15!/2 = 653837184000$ arrangements. And it has been shown that any puzzle, no matter how mixed up it is, can be solved in a maximum of 80 tile moves.

Bonus non-maths but super geeky fact ... American chess superstar Bobby Fischer was a gun at solving the Fifteen Puzzle. He once showed off his skills on the iconic US *Tonight Show* with Johnny Carson on 8 November 1972.

203

Tales from the cryptic

I love a good crossword.

In fact, when I'm not nerding out on numbers I can often be found walking around almost bumping into things because I've got my head buried in the latest cryptic from *The Times* in London.

Yep, I'm quite the cruciverbalist, you could say. I say that not to impress you with my vocabulary, but simply because 'cruciverbalist' is such an awesome word.

There's a horrible rumour[*] going around that once, many years ago, a girlfriend's parents came out the front of the house after we'd all enjoyed dinner together only to find me scrounging in the garbage bin. Not for scraps, but for my cryptic crossword which had been thrown out by an unsuspecting sibling who knew not of my sad addiction.

So imagine my excitement earlier in the year when 5 across in that day's *Times Cryptic* combined my two loves … numbers *and* cruciverbalism:

One-third of twelve still an odd number (6)

I won't spoil the fun for you. The answer is in the back of the book.

[*] I say 'horrible rumour' because I'm to embarrassed to admit it's 100% true. Apologies, Mrs Cassimatis — dinner was lovely.

Here's another mathsy-cryptic clue.

As a hint, it involves an anagram of 9 letters in the clue and the answer is a famous mathematical series that we've spoken about in this book.

Unusually fab, iconic name for a series (9)

Here's one more straight up anagram. No hint. Good Luck.

Tahitian crime out of control — one can control figures (13)

I reached out to one of Australia's most respected cryptologists, puzzle master Iain Johnstone who regularly bends the brains of readers of various Pacific Magazines publications.

So, limber up with these 10 number-based cryptic clues from the mind of the master.

I've supplied hints on page 208, so if you're new to this game, don't be too shy to look for a bit of assistance.

ADAM SPENCER

ACROSS

1. Take the first tune we offer otherwise you can't tango (3)

3. Trafalgar, Tiananmen and SpongeBob's pants need a good day's meals to start working towards 5 (4)

4. Following 4 in the iron (4)

5. Reverse line of symmetry lost a half dozen (3)

6. How many stooges are over there? (5)

7. Rowing crew noisily consumed inbound freighter (5)

DOWN

2. Harry styles old direction found in the money (3)

4. Golfer's call to the boundary (4)

5. First woman runs up between poles to find enough brides for the brothers (5)

6. Half your score strewn evenly across net return (3)

NUMBER LAND

SPOILER ALERT!

Need a hint? Well, I haven't made you flip to the back of the book for this one — just the page. If you want to check your answers before reading the hints, you'll find them at the usual place.

You might have realised that the answers are the numbers one to ten, though not necessarily in that order.

ONE ACROSS
Take the first tune we offer otherwise you can't tango
'Take the first' tells you to look at the initial letters of the words in the clue to find a hidden number.

THREE ACROSS
Trafalgar, Tiananmen and SpongeBob's pants need a good day's meals to start working towards 5
Wow! A beauty here, but tough. What word unites Trafalgar, Tiananmen and SpongeBob's pants?

FOUR ACROSS
Following 4 in the iron
Involves the Roman numeral for 4 and the chemical symbol for iron. Told you this was nerdy.

FIVE ACROSS
Reverse line of symmetry lost a half dozen
What's the mathematical word for a line of symmetry in a shape? Write that word backwards and you're well on your way.

SIX ACROSS
How many stooges are over there?
The second half is anagramming 'there' (using 'over' as a dodgy anagram indicator).

SEVEN ACROSS
Rowing crew noisily consumed inbound freighter
'Inbound' means the answer is contained within another word or words in the clue. Somewhat unconventionally, this clue has a second round of wordplay: a homophone for 'noisily consumed'.

TWO DOWN
Harry styles old direction found in the money
The answer involves taking some letters out of the word of money ...

FOUR DOWN
Golfer's call to the boundary
Think about words used in golf and cricket.

FIVE DOWN
First woman runs up between poles to find enough brides for the brothers
A beautiful clue by my old mate but best you just look up the answer if this one doesn't jump out at you.

    Combine the biblical first woman with the poles of the Earth. Good luck!

SIX DOWN
Half your score evenly strewn across net return
'Return' guides you to read a word or words in the clue backwards to find the numerical answer. Keep an eye out for a second round of wordplay here — 'evenly strewn' points to evenly placed letters in that very phrase.

ADAM SPENCER

Gladiator 2: Gladiator Harder?

The original 2001 blockbuster film *Gladiator* won 5 Oscars, including best picture and actor and took almost half a billion US dollars at the box office.

What's more, it was a great yarn starring Australia — *cough* and New Zealand's — own Russell Crowe.

SPOILER ALERT! If you haven't got around to watching the Oscar-winning classic, maybe go to the next story now and get thee over to Netflix.

All good? Here goes.

When director Ridley Scott contemplated a sequel, he cast his mind back to a moment, just before the end of the film, when our hero Maximus (Crowe) is killed off and pondered a story without the fallen hero at all.

Russell had other ideas — and here's where the story gets really good. Why not commission friend and fellow Aussie legend Nick Cave, who had the screenplay *John Hillcoat's Ghosts ... of the Civil Dead** under his belt, to take a swing at it?

* Nick Cave has been involved with many films that *did* actually get made — usually as the composer of the film's score. IMHO *The Proposition*, *Hell or High Water* and *The Assassination of Jesse James* are three of the very best.

YOU ARE HERE

210

'Hey Russell, didn't you die at the end of *Gladiator 1*?' Cave asked.

'Yeah, you sort that out,' replied Crowe.

Game on.

Cave went on some years later to describe the story on an episode of Marc Maron's aptly named *WTF* podcast.

'So, [Maximus] goes down to purgatory and is sent down by the gods, who are dying in heaven because there's this one god, there's this Christ character down on Earth who is gaining popularity and so the many gods are dying so they send Gladiator back to kill Christ and his followers ... It was a stone cold masterpiece.'

It was also a tough sell.

Crowe's review was reportedly pretty succinct. 'Don't like it, mate.'

All's well that ends well, though, as the great man Shakespeare once said — himself something of a scriptwriting pariah throughout the years.

'I enjoyed writing it very much because I knew on every level that it was never going to get made,' said Cave. 'Let's call it a popcorn dropper.'

Niche tweets

As a public mathematician I accept that some of the media I consume might be considered a little bit ... 'niche'.

Case in point the wonderful Twitter feed @TopologyFact which shares beautiful observations about the mathematical field of topology.

Topologists look at the geometry of a figure and its 'spatial characteristics' as opposed to its precise shape.

Consider the following 5 objects:

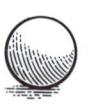

These are all clearly different shapes. But say they were made of plasticine and we were allowed to bend them, mould them, but not push a hole through them or join up bits to fill in a hole or create a new one.

You can easily mould a sphere into a cube and vice versa. But you can't make a sphere into a doughnut (torus) without punching through a hole in the sphere. Similarly, while the plasticine gets pretty thin at the base of the cup, a torus and a coffee cup have exactly one hole through them. But a pair of empty sunglasses has two holes and therefore can't be moulded into any of the other objects while preserving the number of holes.

We say that a sphere and a cube are 'topologically equivalent'. So are a doughnut and a coffee cup. None of those 4 objects is topologically equivalent to the sunglasses.

Another thing topologists might think about is tiling a wall with certain shapes. And this is where Twitter comes in.

You see, one day while sitting around minding my own business — let's be honest, probably solving chess problems online — I read the following:

> *All tessellation types for convex polygons are known, except for pentagons*

What beautiful poetry be this? As is often the case with a statement in mathematics, I knew the meaning of all the words and combining it all together creates an assuredly gorgeous concept but one that slipped from my grasp.

So let's dig in.

A polygon is a shape of many sides. All 3-sided polygons are triangles. They may be big or small, right-angled, isosceles or scalene, but they are all triangles. A square is a 4-sided polygon, as is a parallelogram, a rhombus, a trapezium and other random 4-sided shapes.

We can keep adding sides to get pentagons (5-gons), hexagons, heptagons, octagons, nonagons (yep, that's a word), decagons and beyond.

And in case you're wondering, a 762-sided figure is called a heptahecta-hexaconta-di-gon.

Who am I kidding? You were not wondering that.

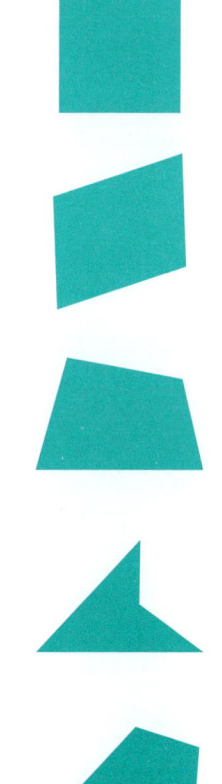

Now, a polygon is 'convex' if 'all of its interior angles are less than 180 degrees'. Ow, that hurt a bit. Well, remember that 180 degrees is a straight line, so for the hexagon below left, all of the angles inside the figure are less than 180 degrees. But for the hexagon on the right we have an indicated a corner (we say vertex) that bends back outwards forming an internal angle greater than 180 degrees. Hence you should be able to see why one is convex and one isn't.

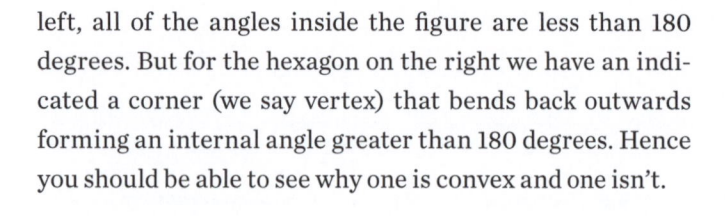

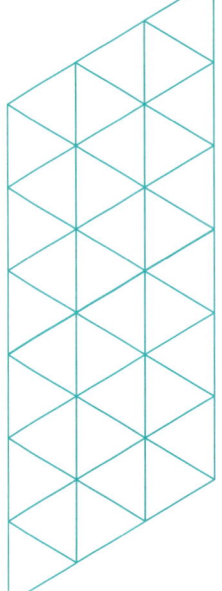

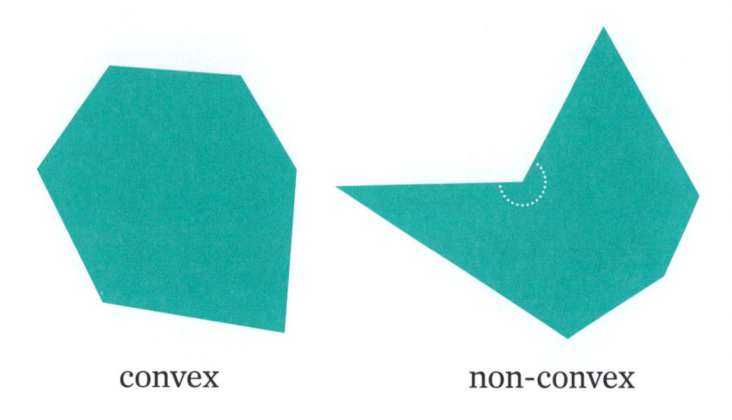

convex   non-convex

We've deciphered the words polygon, pentagon and convex. You could probably draw a convex and non-convex pentagon for me now. But the word that might still elude you is 'tessellation'.

In mathematics, a tessellation refers to covering a plane with a series of shapes that fit together. It may help to explain it is also called a 'tiling'.

At the top of the margin here we see some equilateral triangles (all 3 sides are equal, all angles are 60 degrees) tiling a wall. (Don't worry, I know it's small. I'll give you a gorgeous page of tessellating treats in a minute …)

You should be able to see that this tiling would go on forever. That's if the wall (plane) were infinitely large the tiling could just keep on going and cover it.

We also have an infinite tiling using isosceles triangles (two sides are equal as are two angles), as shown on the bottom left.

In fact, thanks to the work of some awesome topologists, we know that *any* shaped triangle can be used to create an infinite tiling of the plane.

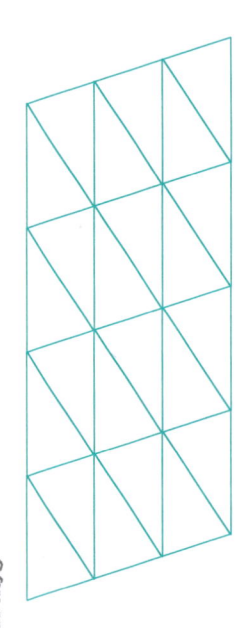

To see this remember that the 3 angles of a triangle add up to 180 degrees — which is also the angle created by a straight line. So any triangle can be placed in the following configuration:

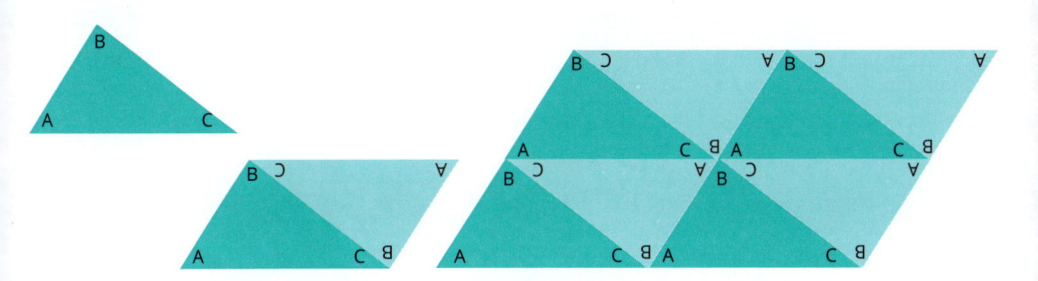

Once you see that you can trace out straight lines as so, you can just extend these rows forever and add them on top of each other, also forever, and cover the entire plane. This works for *any* shaped triangle. Awesome, hey?

What about 4-sided figures, I hear you ask. Well, here's one I prepared earlier.

Take any 4-sided figure (we call it a 'quadrilateral'*) and arrange it this way. Each time you see a circle (as below), you are rotating that shape 180 degrees around that point. Again you should be able to see that the jigsaw pieces fit together in a way that can be extended forever across the plane.

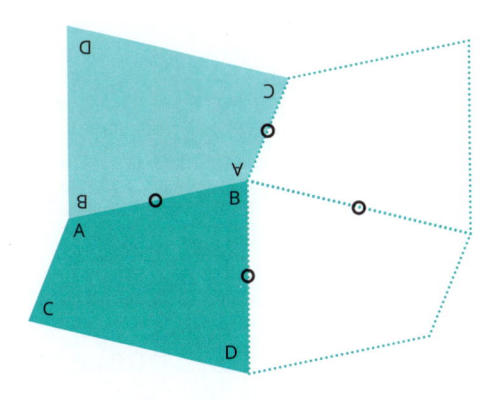

In the case of non-convex quadrilaterals ... off you go and bang ...

* The angle sum of a quadrilateral is 360 degrees.

ADAM SPENCER

YOU ARE HERE

216

You can tile the plane forever.

Before we look at pentagons, let's jump up one step further and consider convex hexagons tiling the plane.

Cometh the hour, cometh the geek. In 1918, German mathematician Karl Reinhardt showed that there are 3 and only 3 such hexagons.

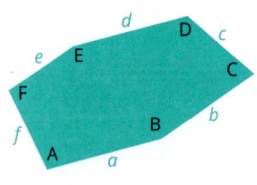

B + C + D = 360°
A + E + F = 360°
a = d

Here is one example of the first hexagon Reinhardt discovered. Note strictly that it is not an individual hexagon but an individual formula for generating a tessellating hexagon. So really there are an infinite number of hexagons that could tessellate the plane because we could tweak the angles B, C and D for example as long as they added up to 360 degrees. But Reinhardt did show his family of 3 types of hexagons covered all possible tessellations of the plane.

In some ways, hexagons are more natural figures to consider when tiling the plane than pentagons. By dividing a hexagon into 4 triangles, we see the internal angles always add up to 720 degrees which is exactly two full circles.

Whereas pentagons can be broken into 3 triangles and therefore have an internal angle sum of 540 degrees: a slightly more inconvenient one and a half circles.

So the most basic regular hexagon clearly tiles the plane, while the pentagon doesn't.

It should come as no surprise that the case of pentagons took a bit longer to crack. But boy were the results beautiful and worth the wait. What's that, you can't wait? Neither can I. Let's dive in.

It was our old German mate the Rein-ster who got the ball rolling. Again in his 1918 PhD thesis he identified 5 pentagons that can tile the plane. But he wasn't sure whether that was all there were. This was already more than we can find for hexagons which, in itself, is exciting.

It took 50 years for Richard Kershner to show, in 1968, 3 more pentagons that tile the plane and to claim that this list of 8 compiled by him and Karl R was all there were.

But he was wrong. In my previous book *Time Machine* (available at adamspencer.com.au) I told the wonderful story of amateur mathematician and homemaker Marjorie Rice who read about pentagonal tilings in *Scientific American* and, armed with only one year of high school mathematics, slogged away — often drawing on pieces of paper on the tiled kitchen benchtop — until she had found ANOTHER 4 PENTAGONS THAT CAN TILE THE PLANE!

This is a beautiful thing. A completely passionate but part-time maths lover adding to the library of human knowledge in a field many experts assumed was closed off for good.

What Marjorie did was to look at the corners of a pentagon and ask herself, 'How could they come together to form a 360-degree sum around a point and could that point be the common vertex of a tiling?' This gave certain rules that the sides and angles of the pentagon would have to follow if they were to tile the plane. You can then ask, 'Is it possible to create a pentagon that satisfies these rules? (Spoiler alert: the answer is 'yes'!)

For example, if a convex pentagon with sides a, b, c, d, e and adjacent angles A, B, C, D, E satisfies the rules:

$b = c = d = e$ and

$2A + C = D + 2E = 360$ degrees

That is, if it looks like this gorgeous mofo on the right.

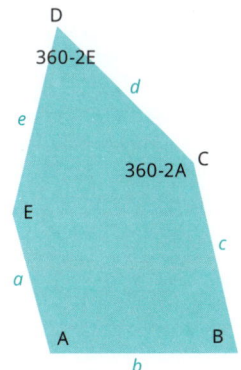

Well, this unit can interlock with itself across the plane ... forever.

Marjorie didn't just find 4 new tiles, she found 60 new tessellations.

In 1975, Richard James III found another tiling, as did Rolf Stein in 1985. A fifteenth example was found by University of Washington. Bothell mathematicians Casey Mann, Jennifer McLoud-Mann and David Von Derau in 2015 using, not a tiled benchtop, but a computer program.

The curtain was pretty much brought down on the search for pentagonal tessellations in 2017 when, with the help of an exhaustive computer search, Michäel Rao showed there are only these 15 examples.

If Rao's paper gets through the peer-review process error-free, we'll know for a fact that all tessellation types are known for all convex polygons, including pentagons. Hey, what an adventure @TopologyFact sent us on.

Just for fun to blow what's left of your mind, hop online and look up the work of Joan Taylor (so close to Joan Tyler, I know) an amateur mathematician from Tasmania who has come up with a single hexagonal tile decorated in a way that it covers an infinite plane without ever repeating itself.

AWESOME.

NUMBER
LAND

YOU ARE HERE

221

The original toll in 1932 to cross the Sydney Harbour Bridge was sixpence a vehicle

It was also threepence for adults and a penny per head for children ... and sheep and pigs.

In the first 24 hours after the Bridge opened, its 25 collectors raised £1500 from about 30,000 cars.

Nice little 'bunse'*.

* 'Bunse' is a shortened form of the (admittedly obscure) Cockney rhyming slang expression 'Bunsen burner, nice little earner' meaning 'a lot of money'. It was popularised by, among others, Ricky Gervais' character David Brent in *The Office*.

223

Origin unknown

If you've heard your kids going off in search for a 'chicken dinner' …

Then it's likely your internet-connected progeny have ignored the Fortnite craze and stayed on 'Murder Island' in PUBG — or 'PlayerUnknown's Battlegrounds' — where a win is announced on screen with the words 'winner, winner chicken dinner'.

While the true origin of the phrase 'winner, winner chicken dinner' may be unknown, there are some fun theories about how it came about.

The fore-running 'origin story' on the interweb seems to suggest that back in the day, a standard bet in a Las Vegas casino was $2 — the same cost as a chicken dinner in those venerable institutions. Thus, a win would return enough to cover … a chicken dinner.

More plausibly, says David Guzman — who quite literally co-wrote the guide on *Craps Lingo from Snake Eyes to Muleteeth*[*] — is that the term originated during the Great Depression.

In this interpretation, people used to play the dice game craps in back alleys to try to rustle up enough money to cover the cost of … a chicken dinner.

Smells a little fishy to me. And I for one am not going to bite into a fishy-smelling chicken dinner. More than likely, Guzman goes on to allow, it could simply have its roots in rhyming slang.

But back to the game. If you haven't played PUBG, you won't be alone — so don't feel too bad. By the same token, if you *have* played it, you're in record-breaking company: for exactly one year up until 9 September 2018, PUBG reached over 1 million concurrent players online every single day. And on the 366th day? 960,263. Oof.

[*] The origin of the term snake eyes is pretty straightforward: the two ones on a pair of die look like … eyes gazing back.

And since it's generally not a great score, the addition of the animal 'snake' creates a nice, anthromoporphic metaphor to the whole thing.

Scores and scores

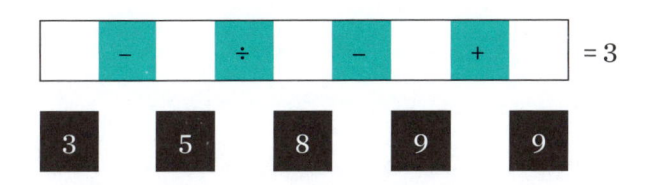

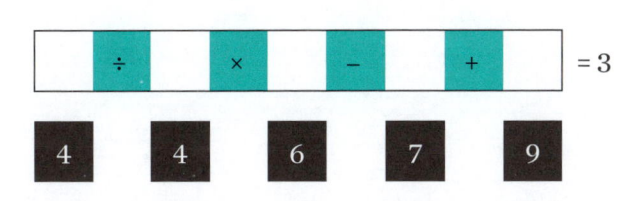

| 3 | − | 5 | ÷ | 8 | − | 9 | + | 9 | = 3 |

| 1 | − | 3 | ÷ | 5 | − | 6 | + | 9 | = 3 |

| 4 | ÷ | 4 | × | 6 | − | 7 | + | 9 | = 3 |

Reach the goal number by creating an equation using the provided numbers and your own choice of operations.

Numbers are placed in white squares, while operations (+, −, ×, ÷) are placed in green squares.

Order of operations matters here, per usual, and you can't use brackets!

Read the numbers off in order for your final score.

You've been given the operators in the correct order. What is the equation that gives the highest score?

225

Sputnik I was our first artificial satellite

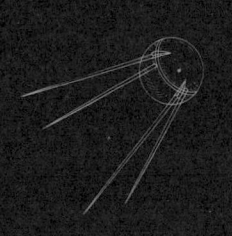

It was fairly small at only 58 centimetres in diameter, but it weighed a hefty 83.6 kilograms.

At a little over 800 kilometres above us (that's less than Sydney to Queensland for comparison's sake) and at 29,000 kilometres per hour, the Russian-designed *Sputnik* completed an orbit of the globe every 96 minutes.

It notched up a total of 1440 orbits in its 3 months of going around in amazing, interstellar circles, before burning up on re-entry on 4 January 1958.

Milky Way's been eating ... too many Milky Ways?

Our galaxy's mass has long been a bone of contention ...

But thanks to NASA's trusty *Hubble* telescope and the European Space Agency's *Gaia* satellite, scientists have determined the Milky Way weighs approximately 1.5 trillion solar masses (in other words 1.5 trillion times the weight of our Sun).

This is considerably heavier than first thought.

And what's all this mass made up of? You might say 'stars' or 'planets' or maybe even 'weird stuff', but you'd be wrong. Partially. Even though there are estimated to be more than 200 billion stars in the MW (and no doubt lots of 'weird stuff'), they represent but a tiny proportion of its mass. Most of the bulk is mysterious, practically undetectable 'dark matter'. In the words of Roeland van der Marel of the Space Telescope Science Institute, dark matter is the 'scaffolding throughout the Universe that keeps the stars in their galaxies'. And compared to other galaxies, the MW is carrying a *lot* of scaffolding.

Exactly how the scientists calculated the MW's mass is a bit beyond the scope of this book, but for the truly keen (I know you're out there), I can tell you they used *Hubble* and *Gaia* to measure the velocities of 'globular star clusters'.

They sound cool to me, too. Globular star clusters (or GSCs for short) are basically isolated space 'islands' that orbit the galaxy, each containing hundreds of thousands (sometimes millions) of stars. Scientists recorded the positions of these stars, creating a sort of 3D map to measure the movement of the GSCs.

When those measurements were combined as anchor points (*Gaia* mapped 34 GSCs and *Hubble* did 12), scientists were able to work out the MW's mass to nearly 1 million light years from Earth. So no, contrary to this story's heading, the MW has definitely *not* been bingeing on chocolate. Though I do like the sound of a 90% cacao treat called Dark Matter...

Take a hike ... on the red planet

Our fascination with Mars (the planet not the chocolate bar) has hit new highs in recent years.

We've successfully landed unmanned spacecraft 8 times (US 6 times; Russia twice) and we first touched down way back in 1976, but NASA's exploration rovers — which were on the ground between January 2004 and June 2018 — enlightened and delighted like perhaps no others.

Scientists are suggesting we may in the not-too-distant future go one further and be able to send humans to the red planet. There are plenty of logistics to sort before that happens — not to mention countless billions of dollars to be spent — but once we get there, I for one, want to know how we'll get about — and what we'll wear. With reduced gravity (Mars has just 38% of the Earth's ... so if you weigh 100 kilograms here, you'll only weigh 38 kilograms there) you'd go slower. According to *Nature* magazine, the optimal walking speed would be 3.4 kilometres per hour, compared to 5.5 kilometres per hour on Earth.

Of course, you'd need adequate clothing, too, as night-time temperatures on Mars go below the freezing temperature of dry ice for many days of the year, while daytime temperatures can get as high as 20° Celsius — but only for a few hours on the warmest days. And because the Martian atmosphere is almost a vacuum, you couldn't survive without an incredibly sophisticated pressurised suit (your saliva and the moisture lining your lungs would boil).

Hmm, I think I'll just stick to a holiday on the Gold Coast.

Digital detox

If you're anything like me, it may be hard to imagine life without a smart device, be it a watch, a phone, a tablet or even a phablet (if they still exist).

But the science is increasingly saying we need to spend less time on our devices — and the stats are stark.

Like gambling machines, smart devices (and more specifically the apps they carry) are designed to trigger the release of the feel-good chemical dopamine in our brains. In other words, repeated use encourages more use and we can quickly become addicted.

But research has shown that smart devices also trigger the release of the stress hormone cortisol. And constantly being stressed is very bad for our health — exacerbating pretty much every chronic disease that we know of, from depression to obesity to high blood pressure to Type 2 diabetes, dementia and stroke.

So what do we do? Simple. Start by turning off all unnecessary notifications. Hide infrequently used apps in folders (or, better, delete them entirely). Take as many breaks away from your device as possible. Try a 24-hour digi detox every once in a while. And be mindful of the cravings for the device and mentally question them.

It ain't easy, but it could improve your life.

Only about 4% of people have an 'outie' belly button

According to a North Carolina State University study of 500 people, that is.

It seems like someone at the uni had a bit of thing for navel-gazing. Another study they conducted (of 60 other belly button-wielding humans) found 2300 forms of bacteria — 1458 of which may even be hitherto unkown forms in the science world.

If belly buttons are your thing, go and check out my mate Dr Karl's IgNoble Award-winning research into why belly button lint is so often blue (spoiler alert: possibly because most clothing contains shades of that colour).

Bean counting

Aussies love a good cup of coffee.

And in the world of coffee snobbery we rank pretty darn well, I'd imagine.

I think it's certainly fair to say we punch above our weight in the quality of the coffee we make.

But in the worldwide rankings of consumption, I'm sorry to say we're simply not pulling our weight. The top 10 looks a little like this:

Finland: 9.6 kg per capita (2.64 cups/day)
Norway: 7.2 kg (1.98 cups/day)
Netherlands: 6.7 kg (1.84 cups/day)
Slovenia: 6.1 kg (1.68 cups/day)
Austria: 5.5 kg (1.51 cups/day)
Serbia: 5.4 kg (1.49 cups/day)
Denmark: 5.3 kg (1.46 cups/day)
Germany: 5.2 kg (1.43 cups/day)
Belgium: 4.9 kg (1.35 cups/day)
Brazil: 4.8 kg (1.32 cups/day)

And Australia?

At just under 3 kilograms per capita, we scrape into the top 50 — at number 42.

> The earliest evidence of coffee being drunk comes from the 15th century in Yemen, however the more modern iteration many now know and love — espresso — was born in Italy when Angelo Moriondo of Turin was granted a patent in 1884 for 'new steam machinery for the economic and instantaneous confection of coffee beverage'.

Knave-gazing in *Numberland*

*N*umberland is a curious place.

Not least because you keep getting stopped by truth-telling knights and lying knaves!

This time, you're approached by 3 different people: Adam, Barrington and Spencer (no relation, I swear).

Adam says, 'Mate, at least one of the following things I'm going to say is true. Trust me. Spencer is a knave. I am a knight.'

At this point, Barrington decides to chime in. 'Adam could claim that I am a knave.'

Spencer laughs and says, 'Neither of those two idiots are knights!'

Oof.

Who is a knight and who is a knave?

> Need a hint? Start off by assuming one of the statements is true and see how the other statements pan out. If you find a dead end ... try a different one.

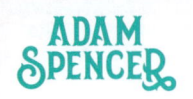

Man vs dog

Next time you're awaiting the bearded barista at your local café to put the finishing flourish on your hamster milk piccolo latte, here's some ponder fodder for you.

A small study published in the February 2019 edition of *European Radiology* suggests that a man's beard may well contain more harmful bacteria that a dog's fur.

The researchers analysed skin and saliva samples from 18 bearded men, ranging in ages from 18 to 76, alongside fur and saliva samples from 30 dogs (ranging from German Shepherds to schnauzers). Though the study was small, the results were not encouraging for our two-legged friends.

The study was not a petty war on hipsters, we're assured, but rather to examine whether it would be safe to use the same MRI scanners for humans and dogs. They tested specifically for human-pathogenic bacteria — the potentially nasty stuff — and found 'high microbial counts' on the skin and saliva of all 18 baristas, I mean bearded

> According to an analysis of US Census data and Simmons National Consumer Survey, some 163 million Americans used disposable razor blades or shavers in 2018.

men. By contrast, 23 of the 30 dogs had minimal counts. Seven of the men tested positive for the human-pathogenic microbes including *Enterococcus faecalis* ... which is pretty much as it sounds — a common gut bacteria.

But before you female readers all go 'ewwwww', the authors also note that 'there is no reason to believe that women may harbour less bacteriological load than bearded men'.

Similarly, the authors highlight that their report — especially given its small sample size — is better viewed as a reminder that we humans leave far more potentially hazardous bacteria around us than we'd care to imagine, even in carefully sanitised places like hospitals.

'The central question should perhaps not be whether we should allow dogs to undergo imaging in our hospitals,' the researchers concluded, 'but rather we should focus on the knowledge and perception of hygiene and understand what poses real danger and risk to our patients.'

No need to reach for the Gillette just yet.

Hubble bubble tile and trouble ...

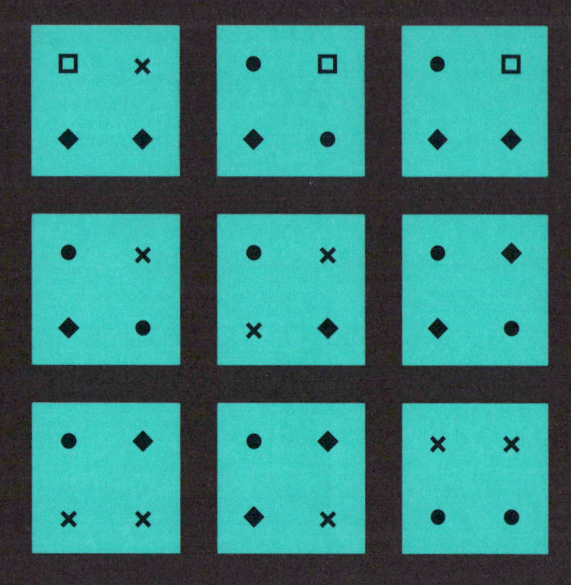

Think you're done tiling the bathroom? Think again, my friend!

Remember to place the 9 tiles into a 3 × 3 grid so that adjacent tiles show the same symbol along their touching edges.

The tiles may need to be rotated, but they are never reflected.

As a tiny clue, the bottom right tiles in each grid are in their correct position, though may need rotating.

Grab the grout.

As always, you can head over to my website to download a cut-uppable copy of this puzzle.

Point your browser to adamspencer.com.au and tile on!

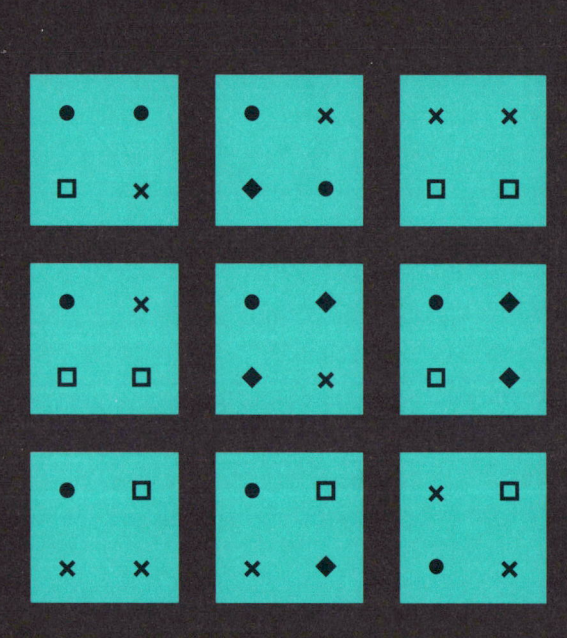

238

Say, this tastes pretty good

I'm beginning to think that a book on the numbers behind mistakes might be worth looking into ...

At Hawksmoor Steakhouse in Manchester, northwest England, managers noticed during a stocktake that they were down a bottle of their very rare 2001 Château le Pin Pomerol — which retails in their establishment for £4500 (roughly AUD$8350). After a little bit of digging, the reason was soon discovered and on 16 May 2019, the restaurant sent out the following tweet:

> *To the customer who accidentally got given a bottle of Château le Pin Pomerol 2001, which is £4500 on our menu, last night — hope you enjoyed your evening! To the member of staff who accidentally gave it away, chin up! One-off mistakes happen and we love you anyway ;-)*

The customer had reportedly ordered the Château Pichon Longueville Comtesse de Lalande 2001 which, at £260 (AUD$480) is no quaffer, but is a mere fraction of the cost of the le Pin.

That's a pretty cool boss. Which reminds me ...

Life goes on ...

One of the most famous examples of a 'cool boss' — and for that matter a great leader — comes from legendary IBM CEO Tom Watson Jr, who led the company between 1956 and 1971.

According to the story, as told in Edgar Schein's book *Organisational Culture and Leadership* and retold in many a business management tome, a young executive made some bad decisions which cost IBM several million bucks.

On entering Watson's office after being summoned for what most people presumed was 'the talk', the employee said, 'I suppose after that set of mistakes you'll want to fire me.'

Watson's reputed response is the stuff of legend. 'Not at all, young man, we have just spent a couple of million dollars educating you.'

The final word on failure should go to a man whose inventions are so legendary he needs no introduction, Thomas Edison, who once quipped, 'I have not failed. I've just found 10,000 ways that won't work.' Similarly, when his factory burned down, along with a large part of his life's work, Edison said, 'There is great value in disaster. All our mistakes are burned up. Thank God we can start anew.'

Well said.

The famous IBM question-answering computer, Watson, was named not after Thomas Watson Jr, but Thomas Watson Sr — his predecessor and, not coincidentally ... his dad.

Incidentally, Watson was initially developed to answer questions on Jeopardy! (see page 196) and in 2011, it competed against legendary champs Brad Rutter and Ken Jennings.

It won first place and a prize of $1 million.

Scores more

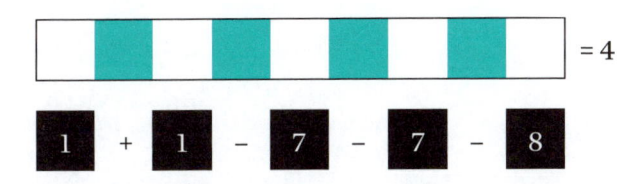

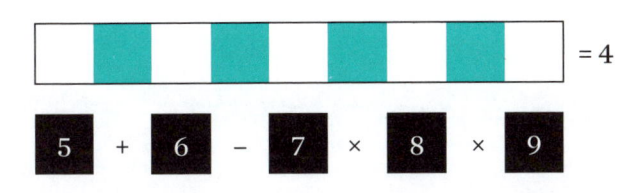

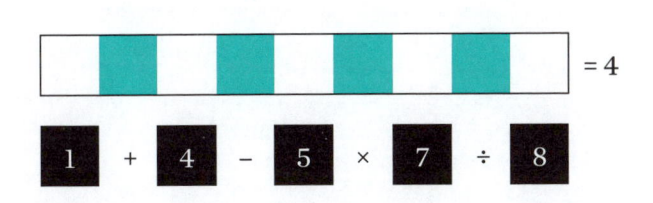

$$1 + 1 - 7 - 7 - 8 = 4$$

$$5 + 6 - 7 \times 8 \times 9 = 4$$

$$1 + 4 - 5 \times 7 \div 8 = 4$$

Reach the goal number by creating an equation using the provided numbers and your own choice of operations.

Numbers are placed in white squares, while operations (+, −, ×, ÷) are placed in green squares.

Order of operations matters here, per usual, and you can't use brackets!

You've been given the numbers and operators for each equation, all jumbled up. Now, work out the order in which to place them!

Read the numbers off in order for your final score.

What is the equation that gives the highest score?

YOU ARE HERE

241

ADAM SPENCER

How long do you reckon dividing lines on roads are?

242

NUMBERLAND

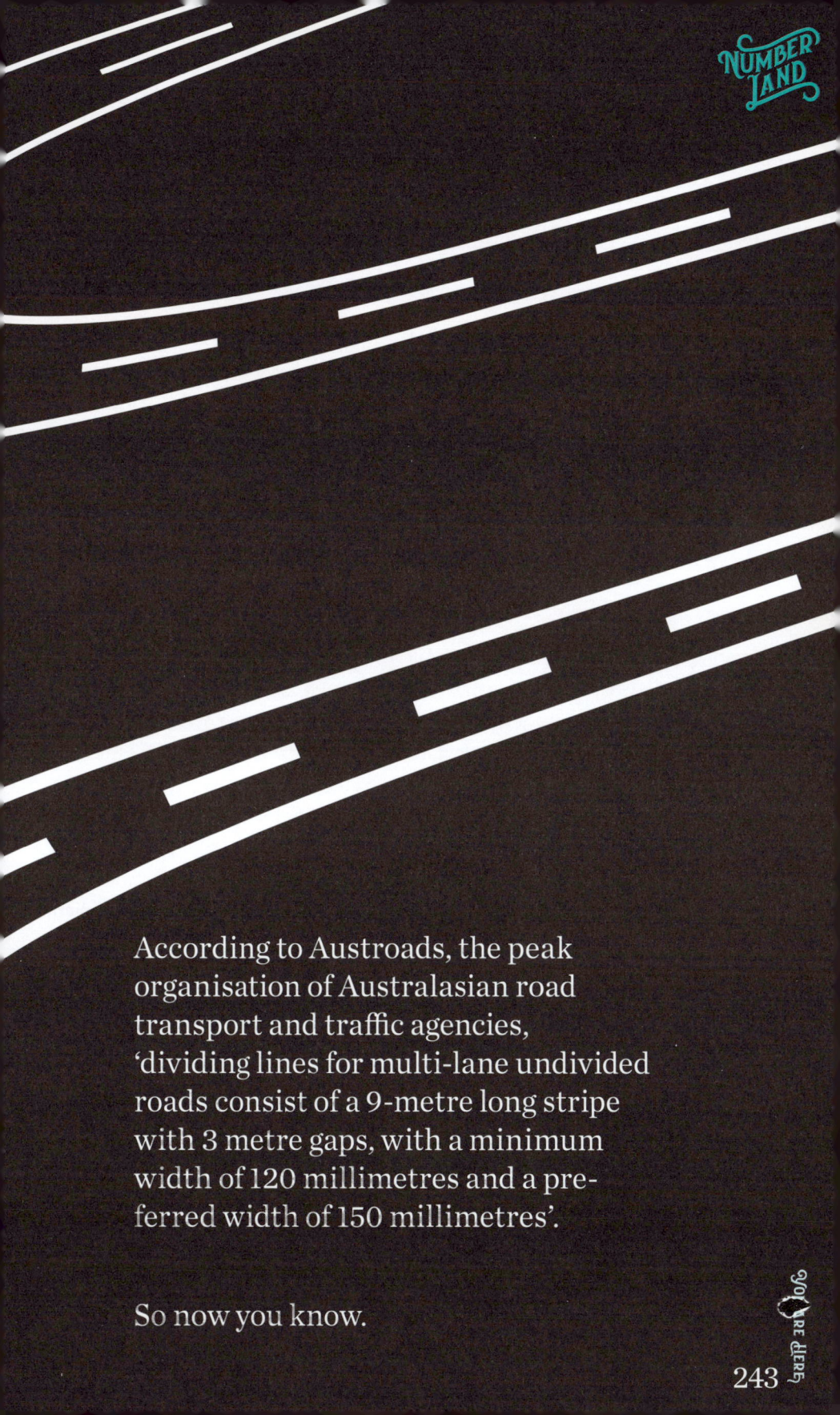

According to Austroads, the peak organisation of Australasian road transport and traffic agencies, 'dividing lines for multi-lane undivided roads consist of a 9-metre long stripe with 3 metre gaps, with a minimum width of 120 millimetres and a preferred width of 150 millimetres'.

So now you know.

21 Jump Heat

In May 2018, the world watched on in horror and fascination as Hawaii's extremely volatile Kilauea volcano went berserk, spewing 320,000 Olympic swimming pools' worth of lava across 35.4 square kilometres of the island.

In doing so, it triggered earthquakes (including one of magnitude 6.9), and destroyed some 700 dwellings.

You'd think that'd be warning enough to tread carefully … and keep a respectful distance. But no. In May 2019, just a year after the massive eruption, a man trying to get a better look at the inside of the volcano climbed over the guardrail at the edge, lost his footing and tumbled 21 metres into the caldera (the volcano's crater).

He fell in around 6.30 pm but wasn't found by Hawaiian fire department workers until around 9 pm. Seriously injured, he had to be airlifted out by helicopter. He was lucky to survive; tragically, in October 2017, someone died while attempting a similar stunt.

The word 'volcano' comes from the Roman god of fire, Vulcan, and they're literally just openings on the Earth's surface. When active — or angry, as some people like to say — volcanoes spew forth ash, gas and magma (which becomes lava once erupted). At 3.2 kilometres wide and 1250 metres high, Kilaeuea is certainly no slouch, but it's a minnow compared to the biggest known volcano in our solar system. That honour belongs to Mars's Olympus Mons which is around 600 kilometres wide and 21 kilometres high.

I wouldn't be jumping *that* guardrail in a hurry.

In case you were wondering, the top 10 countries with the most volcanoes around the world are:

1. United States: 173
2. Russia: 166
3. Indonesia: 139
4. Iceland: 130
5. Japan: 112
6. Chile: 104
7. Ethiopia: 57
8. Papua New Guinea: 53
9. Philippines: 50
10. Mexico: 43

Ronin, Ronin, Ronin ...

On my recent trip to Japan, I was lucky enough to visit Tokyo's magnificent Edo Castle palace.

Built around 1457, the palace has a long, fascinating and frequently bloodthirsty history.

In 1701, in one of the palace's longest and widest corridors, a feudal lord called Asano Takumi-no-kami attacked a high-ranking political foe called Kira Yoshinaka.

Although Kira didn't die, Asano committed ritual suicide out of shame. His supporters, known as ronin (leaderless samurai), then had to avenge their master's honour by killing Kira. In turn, the ronin were obliged to commit ritual suicide out of shame. All 47 of them.

The story has become a powerful symbol in Japan of loyalty, sacrifice, persistence and honour and has been fictionalised on TV, and in Kabuki theatre, novels and puppetry for centuries.

Indeed it has achieved perhaps the ultimate honour in Western culture ... Keanu Reeves made a film about it*.

*Named ... *47 Ronin*. Duh.

Of course there's also the other classic film, *Ronin*, starring Robert De Niro and Jean Reno and featuring one of the greatest movie car chases of all time.

Commuter choreography

I know we have infrastructure issues in Australia — roads that need fixing, train lines that need to be built, airports that need to be upgraded, and so on …

But spare a thought for Japanese rail commuters who each day get around on some of the world's most insanely complex — and congested — networks. All but 6 of the 51 busiest stations on the planet are in Japan. I experienced the chaotic commuter choreography first-hand and it's truly mind-boggling, especially when you consider there are around 127 million Japanese in total living on a land mass of just 377,972 square kilometres. That's a *lot* of humanity in a pretty small area.

Anyway, without further ado, here are the world's top 5 busiest train stations:

- Shinjuku, Tokyo: handling approximately 1,260,000,000 commuters per year
- Shibuya, Tokyo: handling approximately 1,090,000,000 commuters per year
- Ikebukuro, Tokyo: handling approximately 910,000,000 commuters per year
- Umeda, Osaka: handling approximately 820,000,000 commuters per year
- Yokohama, Kanagawa: handling approximately 760,000,000 commuters per year

Amazing numbers, eh? Paris's Gare du Nord is the first station outside of Japan to make the busy list but it only comes in at number 24.

If you were wondering how Australia fares, wonder no more.

Our busiest station is Flinders Street in Melbourne which handles approximately 33,580,000 commuters per year. And to put that into context, our land mass totals a whopping 7.692 million square kilometres and our population is (at the time of writing) just a tick over 25 million. Needless to say, we don't rate a mention on the top 51 busiest train stations list.

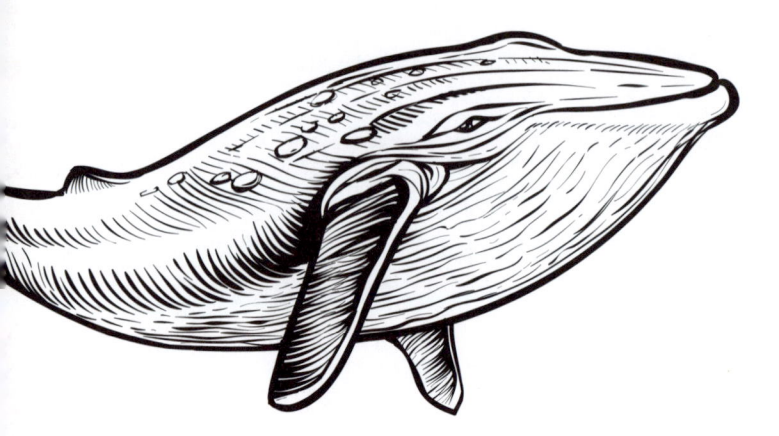

A whale of a time

There's nothing like the discovery of an entirely new species of animal to remind us *Homo sapiens* of how much we still don't know.

Welcome to the party, *Mesoplodon peruvianus*!

This small, grey, 'beaked' whale was first found frolicking in the Pacific Ocean off the coast of Peru back in 1976, but it took scientists another 15 years to verify it was actually a new species.

It's thought that the whales quite sensibly live far from shore and keep away from ships and people — hence their anonymity. They're small too, with adults measuring only around 3 metres in length.

The discovery of the *Mesoplodon* brings the total number of whale species to 13.

At least that we know of today ...

... not such a whale of a time

While we're talking whales, in March 2019, a young Cuvier's beaked whale washed up on a beach in the Philippines, having died from 'gastric shock'.

How did this happen? Well, the whale's stomach contained around 40 kilograms of plastic. The poor creature had ingested countless rice and banana sacks as well as single-use shopping bags — all thanks to humans' carelessness.

Cuvier's beaked whales are fairly common in our oceans but typically hang out in waters deeper than 1000 metres. They grow from 5 to 7 metres in length and can weigh up to 2500 kilograms. They much prefer eating squid and deep-sea fish than plastic bags as they need their energy.

Why? In 2014, scientists off the Californian coast witnessed them diving almost 3000 metres down into the ocean depths. Not only that, they managed to hold their breath and stay under water for over 2 hours. This gives them the honour of being the mammal capable of the deepest and longest dives.

For that achievement alone they surely deserve more than a meal of discarded shopping bags.

Go deep

While our friend the Cuvier's beaked whale may very well be the deepest and longest diving mammal we know of, retired US naval officer, Victor Vescovo, went one better in May 2019.

Aboard his mini-submarine, the *DSV Limiting Factor*, Vescovo travelled to the bottom of the deepest place on Earth — the Mariana Trench (or Challenger Deep, depending on your sources) in the Pacific Ocean east of the Philippines.

And what did he find 10,927 metres down?

No sunlight for one. Freezing water temperatures for two. Water pressure 1000 times greater than at sea level for three. Crazy deep-sea critters like jellyfish, anglerfish, basket stars and acorn worms as well.

But he also found plastic. It's still being tested, but scientists believe it's human's rubbish. I don't know about you, but that doesn't make me feel great about all the packaging I go through in any given day.

The United Nations estimates we've dumped more than 100 million tonnes of the stuff in our oceans to date. People, we need to do better.

Scrabble squabble

Language, like pretty much everything, is in constant evolution.

And if you're anything like me, the barometer for scientifically evaluating linguistic evolution is ... Scrabble, of course! (And the *Macquarie Dictionary*, too, but for the porpoises of this little story, let's stick with Scrabble.)

What words are officially acceptable and what words are not can be the making or breaking of friendships and families. People have been known to get all out of sorts when 'odd' words creep onto the board. Arguments ensue. Voices are raised. Drinks spilled. It can be brutal.

But peeps, I'm here to tell you that circa 2019, Scrabble is now permitting the following leaps forward in the history of the little lettered squares:

sharenting
mansplain
OK
babymoon

Oh, and my personal favourite, *dadbod*.

The old Scrabble rules prohibited anything that was capitalised, foreign, abbreviated or included apostrophes or hyphens. However, the new rules allow proper nouns if they can also be spelled with a lower case initial letter, and foreign words that are commonly used in English.

If you think all that's *ridic*, then *lolz* is all I can say.*

* Yes, those two words were added to the official list in 2015.

The infamous English author, trader and spy best known for penning the novel *Robinson Crusoe* wrote under at least 198 known pseudonyms

NUMBER LAND

Yes, the man sometimes — but best — known as Daniel Defoe also wrote as Jack Indifferent, Timothy Triffle, Tom Manywife, Wallnutshire, Betty Blueskin and my personal favourite, Sir Malcontent Chagrin.

Pure perfection

Let's tie together two ideas we've already touched on in my *Numberland*.

Let's start with idea number 1. Back at page 6, the great Descartes gave props to perfect numbers. What are they?

The factors of 6, apart from 6 itself, are 1, 2 and 3. Add them together and — wow! — 1 + 2 + 3 = 6. This happens very rarely. For 10 we get 1 + 2 + 5 = 8, which is less than 10 and thus called 10 'deficient'. For 12, 1 + 2 + 3 + 4 + 6 = 16 >12, so 12 is called an abundant number. The exact equality that we see in 6 is *so* rare, the Greeks called them perfect numbers.*

Onto idea number 2. On page 42 we met Mersenne primes, primes of the form $2^p - 1$ for p a prime number. Well a cool fact about Mersenne primes is this:

If $2^p - 1$ is prime, then $2^{(p-1)} \times (2^p - 1)$ is perfect.

Don't panic here. Let's use the first few Mersenne primes to generate perfect numbers.

Let $p = 2$. Multiply the Mersenne prime $2^2 - 1$ by $2^{(2-1)}$ we get 3 × 2 = 6, the first perfect number. Similarly when $p = 3$, the formula generates $2^{(3-1)} \times (2^3 - 1)$ = 4 × 7 = 28. Ahoy! Our second perfect number. Convince yourself that $p = 5$ and $p = 7$ generate the next two perfect numbers 496 and 8128.

This means the 24,862,048 digit prime discovered earlier this year leads us to a ... get this ... 49,724,095 digit long perfect number given by $2^{82,589,932} \times (2^{82,589,933} - 1)$.

Now *that's* perfect!

*After 6, the next few perfect numbers are 28, 496 and 8128. If you're up for it, work out all of the factors of these numbers and show that they are perfect!

New high score

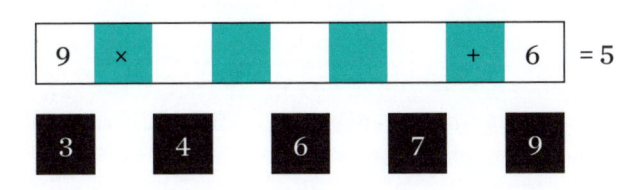

| 9 | − | | | | + | 4 | = 5 |

[3] [4] [6] [6] [9]

| 6 | + | | | | ÷ | 7 | = 5 |

[1] [6] [7] [7] [8]

| 9 | × | | | | + | 6 | = 5 |

[3] [4] [6] [7] [9]

Reach the goal number by creating an equation using the provided numbers and your own choice of operations.

Numbers are placed in white squares, while operations (+, −, ×, ÷) are placed in green squares.

Order of operations matters here, per usual, and you can't use brackets!

The numbers and operators as given in the grids are in the correct position.

Read the numbers off in order for your final score.

What is the equation that gives the highest score?

At the third (and last) stroke ...

In April 2019, it was announced that the watch was going to stop on one of the most iconic pieces of timekeeping Australia has known.

For an impressive 66 years, you could call the number 1194 to find out the precise time. For many people in the digital age this might sound funny. But for generations it was a reassuring and reliable way of knowing when to put the tea on (if all the clocks in your house somehow stopped).

Theatre critic Gordon Gow originally recorded the hours, minutes and seconds in 1953. His dulcet tones were replaced by ABC broadcaster Richard Peach in 1990.

But sadly the service is no longer compatible with modern networks and billing and the 'talking clock' will sound out its last announcements as we tick over from 30 September to 1 October 2019.

Thank you and goodnight.

If you're gonna shoe it, shoe it right*

Australian Formula 1 ace Daniel Ricciardo has made a habit of drinking champagne from his racing boot whenever he gets onto the podium — which, to be fair, isn't that often since he switched teams from Red Bull to Renault ...

But does pounding around an F1 circuit at speeds of up to 350 kilometres per hour for 300-odd kilometres over two or so hours actually produce the ideal vessel from which to sup one's sparkling wine?

Funnily enough it doesn't!

Dr Vincent Ho, from Western Sydney University's School of Medicine Gastroenterology Laboratory, did a little experiment. He added various types of alcohol to well-used sport shoes and tested them for bacteria. While a lot of varieties of booze killed off the baddies — basically disinfecting them — sparkling wine actually encouraged bacterial growth.

Sugar and yeast are added to sparkling wine to produce a secondary fermentation process — this gives it the bubbles — but it also creates a haven for nasties. Particularly harmful varieties such as *Staphylococcus aureus* just love it and thrive. The harmless 'shoey' could quite easily land you in hospital with gastroenteritis.

I'd stick to vodka next time, Daniel. Or just spray the stuff around like everyone else.

*Big thanks to the *Sydney Morning Herald* for that catchy phrase.

Modern Formula 1 cars go fast. Real fast. The quickest ever recorded was clocked by Juan Pablo Montoya's McLaren in testing back in 2005. JP hit an impressive 372.6 kilometres per hour on the straights.

Which certainly is quick. But not compared to the all-time land speed record. That honour belongs to the *Thrust SSC*, a British jet car which hit an astonishing 1228 kilometres per hour back in October 1997, breaking the speed of sound in the process.

Makes an F1 car seem like a family stationwagon!

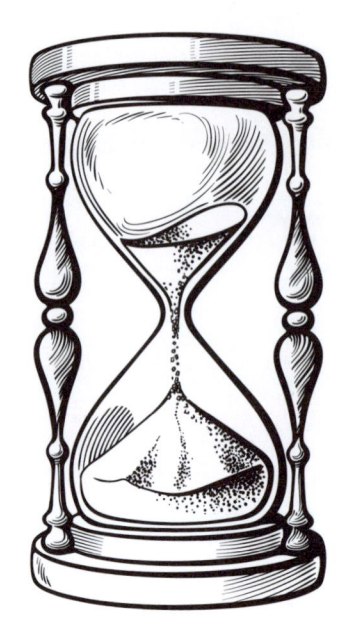

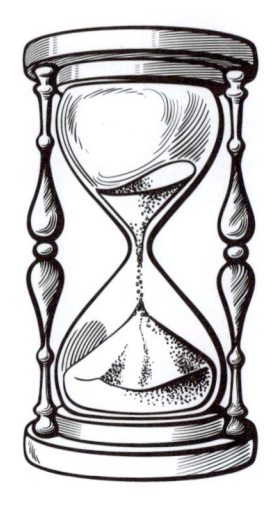

Hourglass half-full kinda guy

This little brain tickler comes from my mate, Reuben Meerman, AKA the Surfing Scientist.

If you're not already following him on all the social medias ... hop to it.

But back to our riddle at hand.

Say you have two hourglasses. You know that one of them empties in 7 minutes. The other empties in 4 minutes.

Can you use these two hourglasses to measure exactly 9 minutes? If so, how?

For whom the bell tolls

Way back in 1799, the British ship HMS *Lutine* foundered off the Dutch coast, resulting in the loss of all but one of her 240 passengers and crew.

The tragedy was enormous, not only for the terrible loss of human life, but because the ship was carrying a fortune in gold and silver — along with the Dutch crown jewels.

You can imagine the underwriters at insurer Lloyd's of London wringing their hands. But in an admirable show of humanity, they paid out the enormous claim in full. Not only that, when the ship's bell was recovered from the depths, it was sent back to London where it was hung from the rostrum of Lloyd's underwriting room.

Traditionally, the bell would be rung when news was received of an overdue ship — one ring for the ship's loss (bad news) or two rings for the ship's safe return (good news). But in these days of the instantaneous 24-hour news cycle, it's more often rung simply to mark special occasions.

Strrrrrrike!

S ports stats fascinate me. Always have. Always will.

Did you know that out of the 9 batters in a baseball team, the 4th batter is known as the 'cleanup hitter' and is always one of the best?

Coaches often place hitters who are likely to get to base just before the cleanup hitter, so that he or she can 'clean up' the bases by driving these baserunners home to score runs.

But for every rule, there's often an alternative. Back in 1956, US baseball manager Bobby Bragan put his best batter in the leadoff position, and the rest of his lineup in descending batting average order.

This lineup turned out to be sheer genius and was mathematically proven so by author Earnshaw* Cook in his 1966 bestseller**, *Percentage Baseball*. While some disputed his maths, Cook claimed that Bragan's lineup resulted in, on average, one to two more wins per season.

Take me out to the ball game!

* Old Earnshaw was a fascinating dude. A mechanical engineering professor, he even worked on the American WWII Manhattan Project (which designed the first nuclear bomb) before becoming a pioneer in sabermetrics — or the science of analysing baseball through statistics.

** The term 'bestseller' is here used advisedly.

Breakfast at ...

Before Charles Lewis Tiffany burst onto the jewellery scene back in 1886, most engagement and wedding rings featured diamonds set into a bezel incorporated into the ring itself so only the top of the stone was visible.

Ol' Charlie thought this was a crying shame as much of the diamond (or other precious stone) was 'buried'.

He and his team proposed a new contraption: a type of raised 'claw' that allowed the diamond to well, shine bright like a diamond.

It was an engineering marvel and so successful that to this day people contemplating the purchase of such things can just request a 'Tiffany setting' and know what they're getting.

And if you were wondering, all Tiffany engagement and wedding rings come in a standard size 6.

A June 2018 study published in *Geochemistry, Geophysics, Geosystems* revealed there may be huge numbers of diamonds lying beneath the Earth's surface.

The scientists said that diamonds aren't all that exotic and are relatively common. There may be as many as 1000 times more than we first thought.

The catch? Most of them are buried hundreds of kilometres down.

Get digging!

Salty stats

The numbers around global nutrition and mortality are truly sobering.

Bad food choices are killing more people than smoking. According to a report funded by the Bill and Melinda Gates Foundation, we're eating way too much salt (and sugar) but not enough whole grains and fruit.

In 2017, bad food choices cut a swathe across the planet. Salty, sugary, processed rubbish contributed to cardiovascular disease (10 million global deaths); cancer (913,000 global deaths) and Type 2 diabetes (339,000 global deaths).

But it's not all doom and gloom. Aussies aren't the worst offenders and we're doing way better than we were 20 years ago. But we've got a long way to go. So stock up on those healthy fruits, nuts, seeds and whole grains … and say 'see ya' to the salt and sugar!

Questacon

In 1988, Australia received some pretty nifty gifts from other countries to celebrate 200 years since Captain Cook's arrival.

Suffice to say our understanding of the significance of that date has become a bit more nuanced since then and we're getting better at celebrating the 40–60,000-odd years of our Indigenous peoples. But I digress.

One cool gift came from the Japanese government — a ¥1 billion gift which was put to good use in building the impressive structure which houses Canberra's Questacon.

If you haven't already been, I can tell you it's well worth the trip for any self-respecting geek. Subtitled the National Science and Technology Centre, it boasts over 200 fascinating (not to mention interactive) exhibits related to science and technology. It's fair to say that numbers make more than a guest appearance, too!

On a recent trip there with my youngest daughter, I came across a fantastic set of brainbusting maths games — a few of which we'll visit here in *Numberland* with the kind permission of the good geeks at Questacon.

For more, head over to their website. But if you have the time, I highly recommend jumping on that bullet train* and making your way to King Edward Terrace in Canberra to visit the place in person.

*Look, I've got to be honest, you may be waiting on that train.

As a few wags on Twitter noted, it's scheduled to run fairly infrequently: once every Federal election, to be precise ;-)

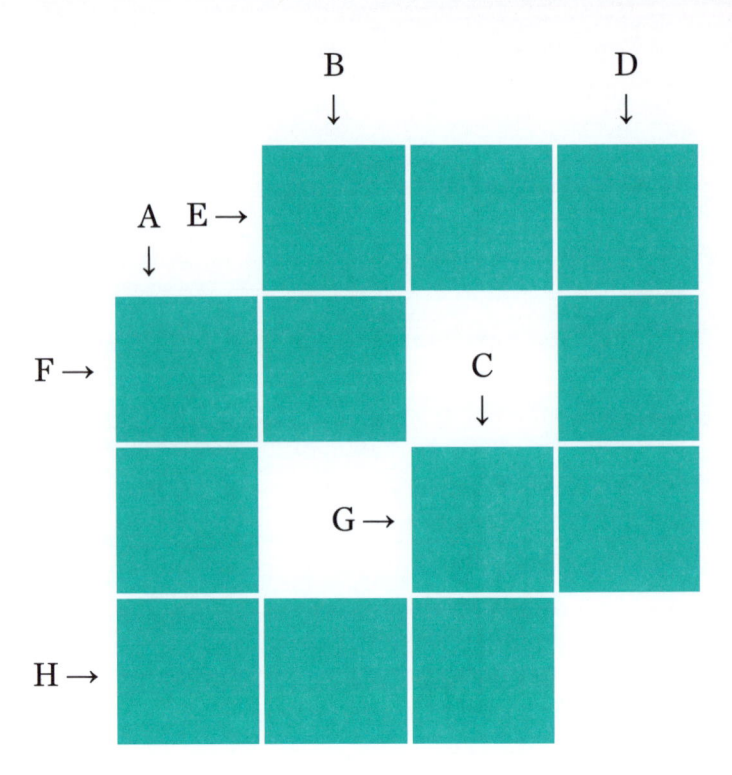

Questacontent 1

As the good folk at Questacon behind these mathematics games say, 'Perplex your mind with calculating conundrums, genius geometry and tricky trigonometry' ... along with a healthy dose of alliteration!

This puzzle is similar to a crossword. Solve the clues and write the numbers in their correct places.

A A multiple of 11
B and **F** Two primes, which when added make 54.
C + 100 = **H**
D is one of the 3 square numbers on the board
E is one of the two cube numbers on the board
G is a square number *and* a cube

Greyscale

Amid the *Game of Thrones* furore that seemed to sweep much of the planet in 2019, 3 small words would have sent chills down the spine of actor Jacob Anderson.

'Go to prosthetics!'

Such an order ordinarily would have meant his character, Grey Worm, leader of the Unsullied, was required to spend time being made up to look the part on screen after some untimely violence ... or worse.

Happily, GW survived and he and fellow actor Ian McElhinney (who plays Ser Barriston Selmy) got a farewell party instead.

Winter, averted. For some, at least!

Over its 8 season run, *Game of Thrones* has used a breathtaking 15,141 litres of fake blood — some of which (SPOILER ALERT!) would have been put to good use in the scene depicting Ser Barriston's demise as he was stabbed multiple times.

Masterstrokes

In April 2019, golfer Tiger* Woods won the US Masters for the 5th time.

It was his 15th major title and was all the more remarkable as it was his first major win in 11 years following a series of, *ahem*, personal setbacks, including spinal surgery.

Woods' achievements are impressive, but he's not the only golfer to clock up some cool sporting stats at the famous tournament held each year in Augusta, Georgia.

Let's take a quick run through my top 10:

10. **63** The course record
    (shared by Nick Price and our own Greg Norman)
9. **50** The most consecutive starts made by one player
    (Arnold Palmer, 1955–2004)
8. **46** The age of the oldest winner (Jack Nicklaus, 1986)
7. **21** The age of the youngest winner
    (Tiger Woods, 1997)
6. **18** The lowest total under par
    (Tiger Woods, 1997; Jordan Spieth, 2015)
5. **12** The widest margin of victory (Tiger Woods, 1997)
4. **11** The most birdies in a single round
    (Anthony Kim, 2009)
3. **8** The best comeback after 54 holes (Jack Burke, 1956)
2. **7** The most consecutive birdies
    (Steve Pate, 1999; Tiger Woods, 2005)
1. **6** The most number of victories
    (Jack Nicklaus, 1963, 1965, 1966, 1972, 1975, 1986)

Okay, looking at those numbers, Tiger does feature pretty heavily. Played.

---

* Eldrick.

Yep, Tiger's full name is Eldrick Tont Woods. The name 'Tiger' is actually a nickname given by his father Earl at a very young age.

Suffice to say, if you stood on the fairway screaming 'Go Tont' as Tiger swatted one, you'd likely get some odd looks.

'When you're concentrating hard, hours can fly by, and it's just you and a math problem'.

—Terence Tao

This is not a spiral

If you spend any time on the internet or indeed have read my previous books, you'll be familiar with the astonishing beauty of Akiyoshi Kitaoka's work.

Akiyoshi is a professor of psychology at Ritsumeikan University in Kyoto, Japan and specialises in making wonderful, dizzying optical illusions by playing with geometry, colour and brightness. Just like the 'spiral' you see opposite. Or do you? In fact, what you are looking at is actually a series of concentric circles.

This is Akiyoshi's take on the 'twisted cord' illusion, first described by the British psychologist Sir James Fraser in 1908. The distortion is produced by combining a regular line pattern, such as the circles, with misaligned parts — in this instance the triangular segments.

You can verify the concentric strands manually by tracing one of the circles. It should then become obvious that there's no spiral at all ... and the illusion disappears.

Check out Akiyoshi's spectacular website at www.ritsumei.ac.jp/~akitaoka/index-e.html — though be warned: it 'contains some works of "anomalous motion illusion", which might make sensitive observers dizzy or sick. Should you feel dizzy, you had better leave this page immediately!'

Image courtesy of the brilliant Akiyoshi Kitaoka, 2019

ADAM SPENCER

Pandas Eat Milk Duds And Skittles

Or, should I say, 'Please Excuse My Dear Aunt Sally', (she goes caving with Nelly's Nanny, after all ... you'll meet her later on ...)

Despite their absurdity, these nifty *Numberland* mnemonics help remind us of the trusty order of operations: Parentheses, Exponents, Multiply, Divide, Add and Subtract.

You may have remembered the same law with the word BIDMAS where B and I represented Brackets (Parentheses) and Indices (Exponents).

Me? I think the Skittle-eating panda is a tad cuter.

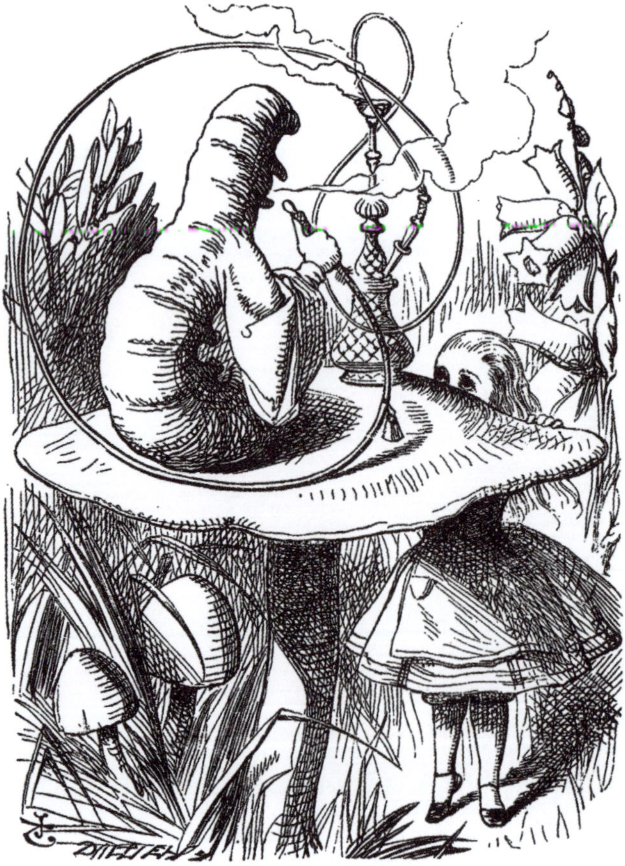

The Caterpillar

Lewis Carroll's 3-inch-tall grub is one of his most intriguing characters — not least because of his love of flavoured tobacco.

The pedantic, persistent little fella has been likened to a Victorian-era school principal ... while the 1960s and 70s 'flower power' generation claimed him as a bona fide 'stoner'. I just think he was misunderstood.

Real life caterpillars (order *Lepidoptera*) are sometimes dismissed as (very) hungry agricultural pests. They're mainly vegetarian — and yes, they are basically eating machines — but some eat insects and some eat other caterpillars. Consider their lifespan: they only live a few weeks before building their cocoon and hatching a few weeks after that ... with wings! And although some are brightly coloured, poisonous and look fierce, birds and other insects just love snacking on them.

What are some other fascinating caterpillar facts? Glad you asked. There are more than 20,000 species around the world. They increase their body mass as much as 1000 times or more during their short life. A monarch caterpillar, for example, is capable of eating one whole milkweed leaf (up to around 25 centimetres in length) in under 4 minutes. They have up to 4000 muscles. And they have 6 legs (and many more 'false' prolegs) and 12 eyes (but terrible eyesight).

Caterpillars cause a surprising number of stings for such small creatures. A lot of them have hairs or spines and when these come into contact with human skin, they can cause pain, rashes, itching, burning, swelling and blistering. Best advice? Give them a wide berth. If you do happen to get stung, try getting the spines or hairs out of your skin using sticky-tape. And if pain persists? Well, my friend, has television advertising taught you nothing? See your doctor or pharmacist.

Illustration by John Tenniel

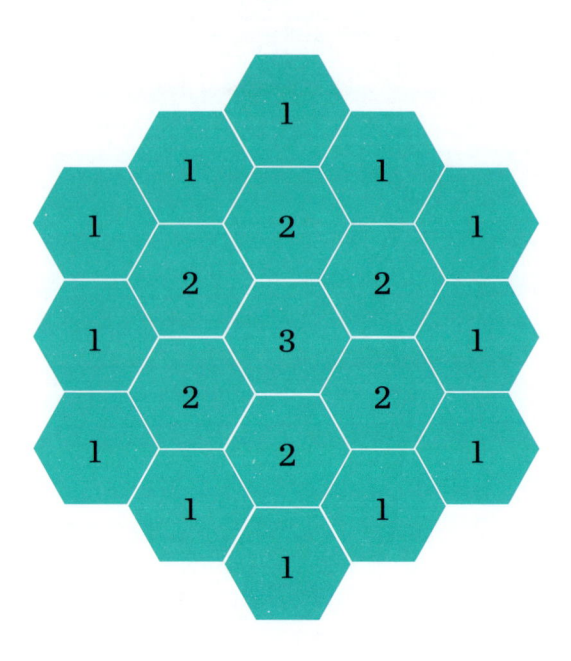

Questacontent 2

Here's another calculating conundrum (with a measure of genius geometry) and some obligatory alliteration from our friends at Questacon.

Moving from one hexagon to another, how many paths can you find that count out 1, 2, 3?
    This puzzle should limber you up for a larger hexagonal maze we'll meet further down the path in *Numberland*.
    Tackling this one first will help you on your quest.

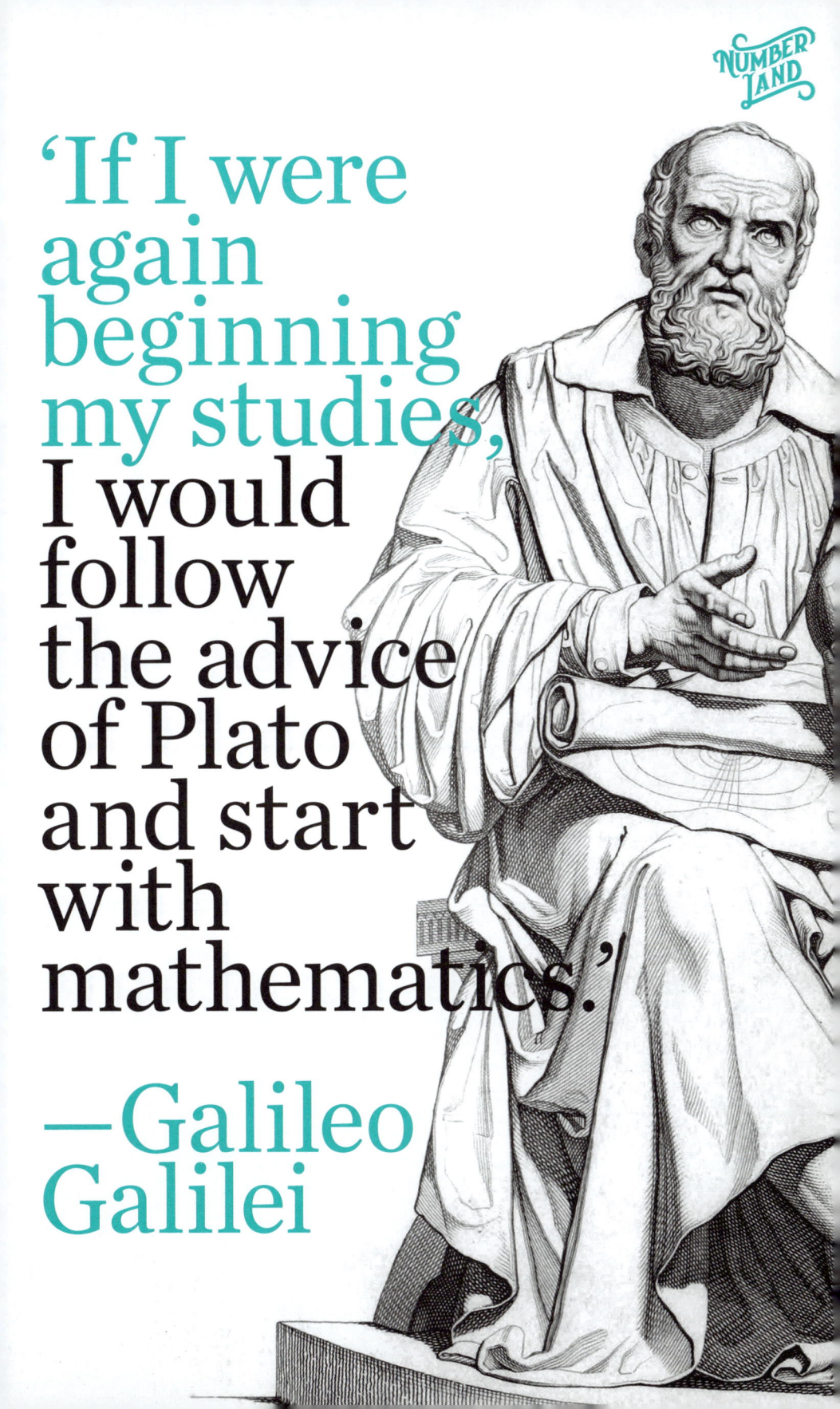

NUMBER LAND

'If I were again beginning my studies, I would follow the advice of Plato and start with mathematics.'

—Galileo Galilei

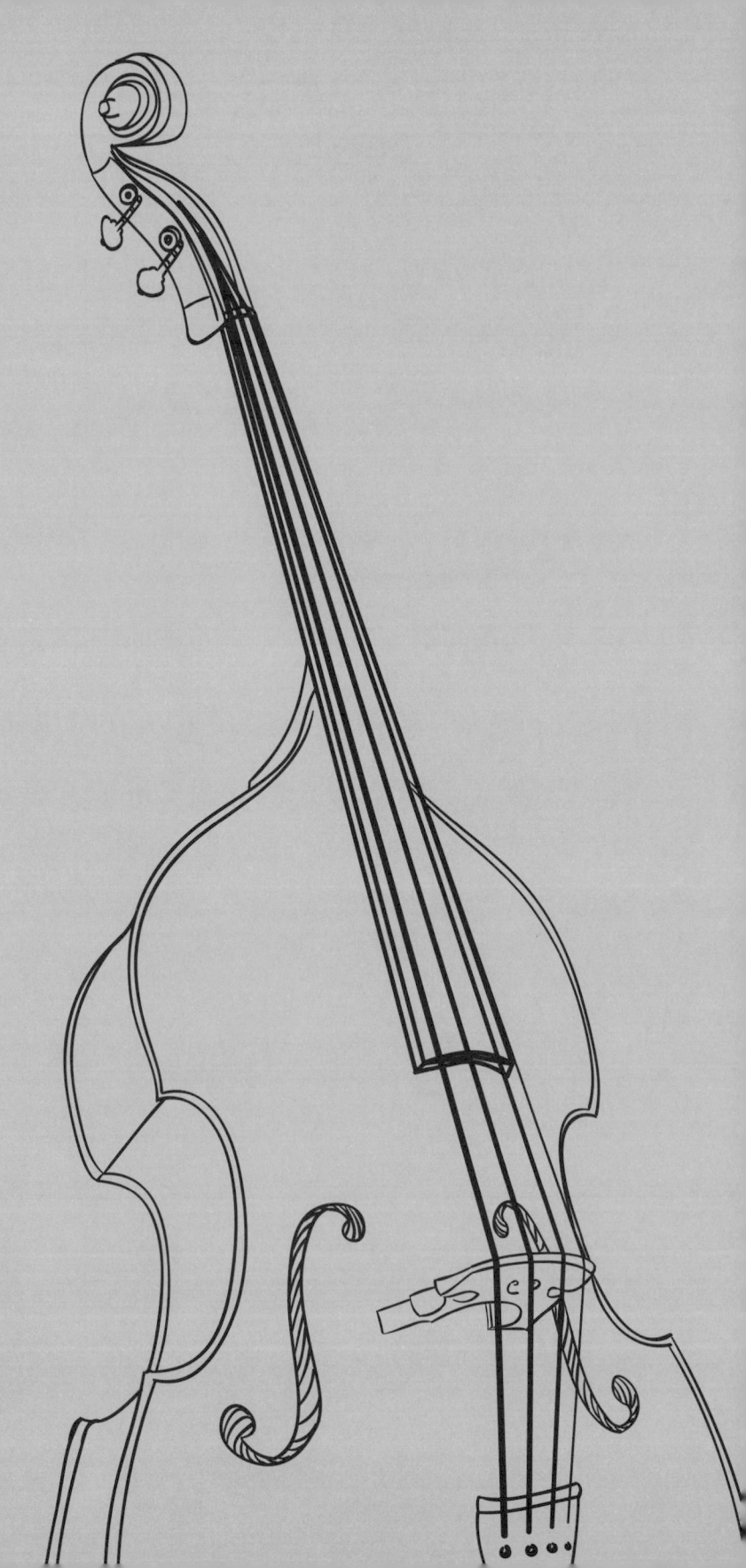

ADAM
SPENCER

276

There are 4 strings on a cello

Now, this is a perfectly interesting fact, but I wouldn't have thought it was AMAZING.

Shows what I know. Whenever I type 'number of' into Google, this is the first things it auto-suggests. So I guess it's a fairly burning question among fellow Aussie number Googlers.

Anyway, cellos have 4 strings tuned in perfect fifths, much like a violin. The notes are C, G, D and A, in descending order of thickness.

Perhaps you could remember this by a new *Numberland* mnemonic — Cellos Glide Delightfully, Adam.

There you go. #savedyouaclick

ADAM SPENCER

'Pure mathematics is, in its way, the poetry of logical ideas.'

—Albert Einstein

Photo of Einstein in 1921 by F Schmutzer

The knave, the knight ... and the footy player?

And so we once again cross paths with *Numberland*'s knightish knaves. Or are they knavish knights?

Our trio this time are easily recognised: one wears a Collingwood jersey, one wears an Essendon jersey and the other (legend!) wears a Sydney jersey. We know that one is a knight, one is a knave and the other is a footy player.

'Which one of you guys plays footy?' I ask.

The man wearing the Collingwood jersey says, 'I can tell you that the man in the Essendon jersey is a footy player.'

The man in the Essendon jersey says, 'No, mate. The girl in the Sydney jersey is a footy player.'

The girl in the Sydney jersey rolls her eyes and says, 'Nope. The guy in the Essendon jersey is definitely the footy player.'

We check our *Numberland* guide book and discern that football players are allowed to lie or tell the truth as they please.

Armed with this fact, who is the footy player? Who is the knight? And who is the knave?

HINT: Two of our merry trio reckon the Essendon jersey-wearer is the footy player. What does that suggest?

The nerdy *Numberland* half-century

You've heard of Triple J's iconic *Hottest 100* …

Well, let me welcome you to *Numberland*'s nerdy *Half-Century*.

One of the most famous musical charts, *Billboard* magazine's 'Hot 100' first featured in its pages on 4 August 1958. Until 1991, it determined its rankings based on playlists reported by US radio stations and surveys of retail outlets. On 30 November 1991, *Billboard*'s chart went high-tech and collated data from a range of sources — not least of which is Nielsen Broadcast Data Systems and SoundScan.

And my list? I'll leave you to discern how these chart-toppers were determined, but suffice to say it's a little nerdier and far more arbitrary. Oh, and you'll notice the occasional tie. I'm not going to lie. Some of the tracks on this list are abominable earworms, and I can't say I wholly endorse too many lists which put Britney at number 1, however that's how it is in *Numberland*, folks.

So, without further ado, here we go.

It's amazing how the way we consume music has changed in my lifetime. Until 1998, to qualify for the Billboard Hot 100 you had to release your track as a physical single.

#	SONG	ARTIST
1	... Baby One More Time	Britney Spears
2	Song 2	Blur
3	Three Little Birds	Bob Marley & The Wailers
4	Four Seasons In One Day Positively 4th Street	Crowded House Bob Dylan
5	It's Five O'Clock Somewhere	Jimmy Buffett & Alan Jackson
6	6 Inch	Beyoncé (ft. The Weeknd)
7	Seven Nation Army	The White Stripes
8	Eight Days A Week	The Beatles
9	Cloud Nine	The Temptations
10	10 X 10 10 Seconds Down	Yeah Yeah Yeahs Sugar Ray
11	The Eleven Jam	Grateful Dead
12	12 Bar	UB40
13	No. 13 Baby Conversations With My 13 Year Old Self Thirteen Women (And Only One Man In Town)	Pixies Pink Bill Haley & His Comets
14	March 14	Drake
15	5:15 15 Step	The Who Radiohead
16	Sixteen Saltines Sixteen (Pop) Only Sixteen	Jack White Iggy Pop Sam Cooke
17	17 Again	Eurythmics
18	I'm Eighteen	Alice Cooper
19	19th Nervous Breakdown	The Rolling Stones
20	Twenty Years	Placebo
21	21 Guns 21st Century Man	Green Day Electric Light Orchestra
22	22 22 Two's	Taylor Swift Jay-Z
23	23	Blonde Redhead
24	24 Hour Party People	Happy Mondays
25	Twenty-Five Miles	Edwin Starr

26	00:26	Ólafur Arnalds & Nils Frahm
27	27 And Single	Joan Rivers
28	Opus 28 28° C Twenty Eight	Dustin O'Halloran Takashi Wada The Weeknd
29	Highway 29	Bruce Springsteen
30	16 Shells From A Thirty-Ought-Six	Tom Waits
31	31 Seconds Alone	Ke$ha
32	The 32nd of December 32 Fahrenheit 32 Footsteps	Babyshambles Jane Birkin They Might Be Giants
33	The Pilgrim: Chapter 33 (Hang In, Hopper)	Kris Kristofferson
34	White Of The Eye (Parts 1–34)	Pink Floyd
35	Rainy Day Women #12 & 35	Bob Dylan
36	36 Degrees 1.36	Placebo Coldplay
37	Rabbits Are Roadkill On Route 37 California 37	AFI Train
38	38 Years Old 38.45 (A Thievery Number)	The Tragically Hip Thievery Corporation
39	Pier 39 39 Thieves	Bon Iver Aesop Rock
40	40 A Pirate Looks At Forty 40 Hour Week (For A Livin')	U2 Jimmy Buffett Alabama
41	Forty One Mosquitos Flying In Formation 41st Street Lonely Hearts Club	Tame Impala Buck Owens
42	Miracle On 42nd Street	The Flaming Lips
43	Hymn 43	Jethro Tull
44	.44 Caliber Love Letter 4:44	Alexisonfire Jay-Z
45	45:33 45th Floor	LCD Soundsystem The Doobie Brothers
46	Forty Six & 2	Tool
47	4:47 AM (The Remains Of Our Love)	Roger Waters
48	48 Crash 48 Hours 48	Suzi Quatro The Clash Tyler the Creator
49	49 Bye-Byes	Crosby, Stills, Nash & Young
50	50 Ways To Leave Your Lover	Simon & Garfunkel

Questacontent 3

Time to 'perplex your mind with calculating conundrums, genius geometry and tricky trigonometry' (not to mention your daily dose of alliteration) with more awesome Questacontent.

Beginning at the top-left corner, try to trace a route in this unusual maze. Move in the direction of the arrows, in distances cycling from 1 square, then 2 squares, then 3 squares, then 4 squares, then 1 square, then 2 squares, and so on.

If you fall off the grid ... jump back on and start again.

… Harry, He Likes Beer Bottles Cold, Not Over Frothy. Nelly's Nanny Might, Although (Silly Person) She Climbs Around Kinky Caves …

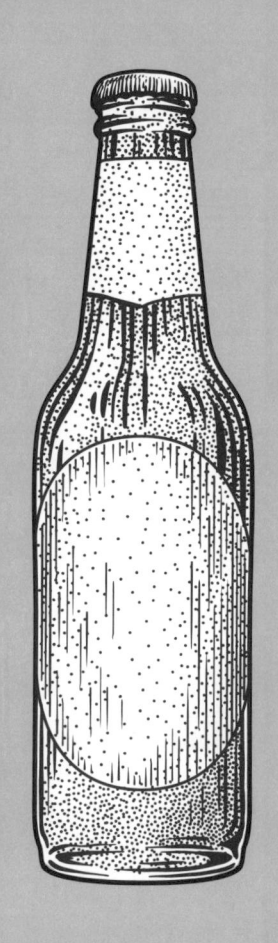

Look, whatever floats your boat, Nelly's Nanny. I'm not going to kink-shame your speleological proclivities.

Nevertheless, you might find this little piece of gossip useful if you're trying to remember the first 20 elements of the periodic table: Hydrogen, Helium, Lithium, Beryllium, Boron, Carbon, Nitrogen, Oxygen, Fluorine, Neon, Na (Sodium), Magnesium, Aluminium, Silicon, Phosphorus, Sulphur, Chlorine, Argon, K (Potassium) and Calcium.

Score'n

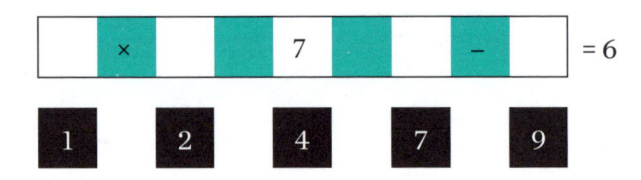

```
_ × _ 7 _ _ = 6
1  2  4  7  9
```

Reach the goal number by creating an equation using the provided numbers and your own choice of operations.

Numbers are placed in white squares, while operations (+, −, ×, ÷) are placed in green squares.

```
_ + _ 3 _ × _ = 6
3  5  6  9  9
```

Order of operations matters here, per usual, and you can't use brackets!

Again, any numbers and operators in the grid are in their correct positions.

Read the numbers off in order for your final score.

What is the equation that gives the highest score?

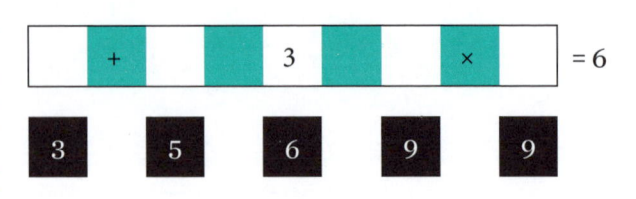

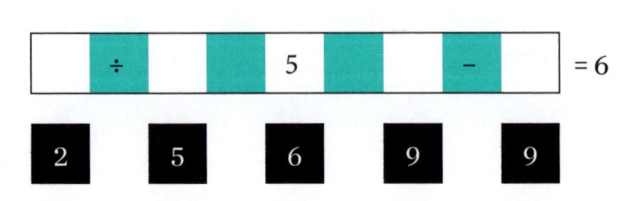

```
_ ÷ _ 5 _ − _ = 6
2  5  6  9  9
```

Twins ...
with a twist

A twin is one of two children (or animals) born at the same birth (that is, by the same mother).

I probably haven't added much to your pub trivia arsenal with that. But wait — hold my beer.

In Australia, twins account for about 1 in 80 births, and there are, most commonly, two different types: monozygotic (MZ), or identical; and dizygotic (DZ), or fraternal or non-identical.

DZ twins occur when two eggs are released at a single ovulation and are fertilised by two different sperm. This is the most common type — 4 in 1000 in Asian populations, 8 in 1000 in Caucasian populations, and 16 in 1000 in African populations. This is interesting since it suggests a genetic component to DZ 'twinning'. They share the same type of genetic relationship as their more garden-variety non-twin siblings.

MZ twins develop when, during the first two weeks after conception, an egg is fertilised by a single sperm, but the developing embryo splits into two. Here, two genetically identical babies develop. The rate of MZ twins is consistent across the globe — about 4 in 1000 births — which suggests a random biological phenomenon.

But researchers at the University of New South Wales and the Queensland University of Technology recently identified a pair of semi-identical, or *sesquizygotic* twins. This extremely rare phenomenon occurs when *two sperm* fertilise the *same* egg.

The 4-year-old boy and girl are from Brisbane and, in case you're wondering how their stats stack up ... they're only the second known pair in the world.

According to the Kübler-Ross model, there are 5 stages of grief

They were first introduced in Swiss-American psychiatrist Elizabeth Kübler-Ross's 1969 book *On Death and Dying*.

The stages are popularly known by their acronym DABDA and include Denial, Anger, Bargaining, Depression and, finally, Acceptance.

Though they are popularly repeated, including in the sitcom *Frasier* and even *The Simpsons*, there's little objective clinical research to back them up.

'I couldn't afford to learn it,' said the Mock Turtle with a sigh. 'I only took the regular course.'

'What was that?' inquired Alice.

'Reeling and Writhing, of course, to begin with,' the Mock Turtle replied; 'and then the different branches of Arithmetic — Ambition, Distraction, Uglification and Derision.'

'I never heard of "Uglification",' Alice ventured to say. 'What is it?'

The Gryphon lifted up both its paws in surprise. 'What! Never heard of uglifying!' it exclaimed. 'You know what to beautify is, I suppose?'

'Yes,' said Alice doubtfully, 'it means —to — make — anything — prettier.'

'Well, then,' the Gryphon went on, 'if you don't know what to uglify is, you *are* a simpleton.'

— from *Alice's Adventures in Wonderland*

NUMBER
LAND

YOU ARE HERE

291

Hard score

```
[ ] [×] [ ] [ ] [7] [ ] = 7
 2   5   7   9   9
```

Reach the goal number by creating an equation using the provided numbers and your own choice of operations.

Numbers are placed in white squares, while operations (+, −, ×, ÷) are placed in green squares.

```
[ ] [ ] [÷] [5] [ ] [ ] [ ] = 7
 1   4   5   5   6
```

Order of operations matters here, per usual, and you can't use brackets!

Read the numbers off in order for your final score.

As in the previous puzzles, numbers and operators in the grids are in their correct positions. Enjoy these clues — after this, you're on your own!

What is the equation that gives the highest score?

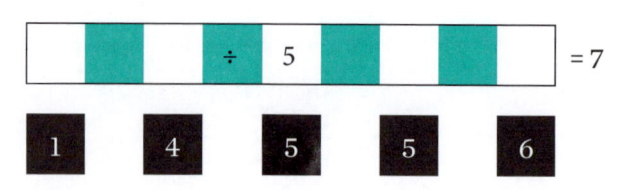

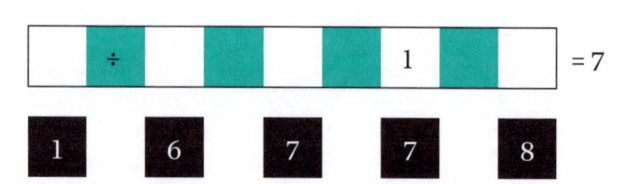

```
[ ] [÷] [ ] [ ] [1] [ ] = 7
 1   6   7   7   8
```

292

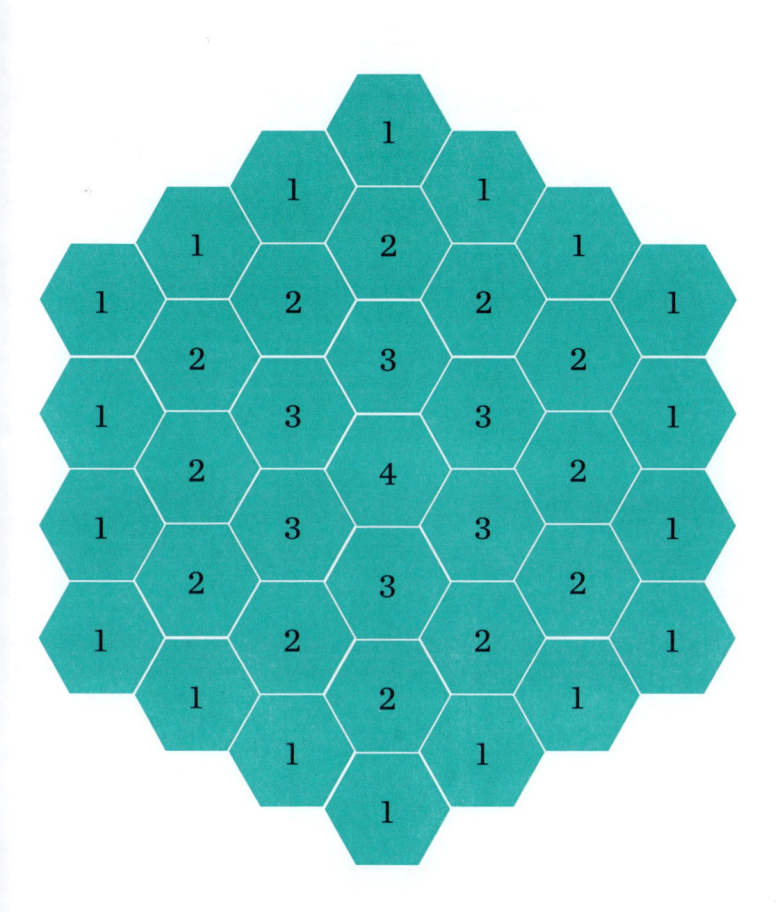

Questacontent 4

O nce again we find ourselves at the start of an intriguing hexagonal maze courtesy of the fiendish riddlers at Questacon.

Moving from one hexagon to another, tell me how many paths you can find that count out 1, 2, 3, 4?
   Get to it!

Tackling the first, smaller version of the puzzle back on page 274 might just help you solve this second challenge …

you are here

293

ADAM SPENCER

The Cheshire Cat

Lewis Carroll's grinning feline is one of the most memorable and endearing characters in *Alice's Adventures in Wonderland*.

Witty and wise, capable of disappearing when in danger (leaving only a smile) he pre-dates his more modern relatives the lolcat, the cat in the hat, Crookshanks, and the cartoonish Tom, Garfield and Sylvester.

In honour of our smiley, semi-invisible friend, here are some feline facts to fascinate you.

It's thought the ancient Egyptians domesticated cats some 4000 years ago. DNA evidence shows that modern-day kitties share a common ancestor with the Middle Eastern or African wildcat so this theory holds water. But in 2004 a 9500-year-old grave in Cyprus was discovered. Inside, archeologists found the remains of a cat buried alongside a person. But some scientists think we befriended these fur balls some 12,000 years ago. Point being, we share a long history together.

Cats don't really 'smile'. They have muscles in their faces that make expressions that resemble smiling, but it has nothing to do with happiness. In fact, if you come across a cat baring its teeth (and hissing) it's probably far from happy. A content cat will purr, blink slowly, knead its paws, rub its head and meow. But it won't grin.

Guinness World Records have stopped recognising the fattest pets in the world because people were overfeeding them just to break a record but ... back in 2003, a 5-year-old Siamese named Katy from Asbest in Russia had to be given hormones to stop her breeding. Unintended consequences of the hormones included an insatiable appetite. Katy packed it on and tipped the scales at a whopping 23 kilograms. About as heavy as a 5-year-old human!

Illustration by John Tenniel

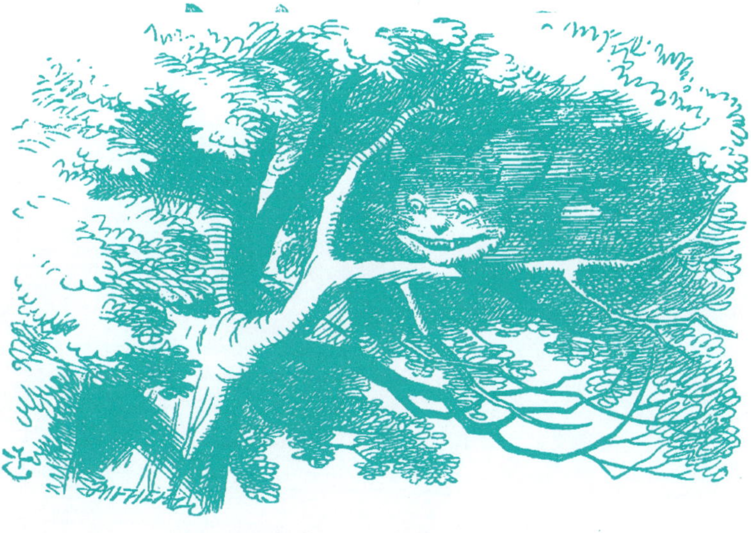

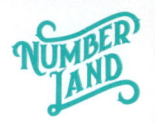

I Value Xylophones Like Cows Dig Milk

It's totally true.

But it's also a handy mnemonic device to remember ... the order of Roman numerals: I = 1, V = 5, X = 10, L = 50, C = 100, D = 500 and M = 1000.

Et tu, Daisy?

Clever Hans was a fraud

He couldn't, after all, calculate the answer to mathematical problems.

That's not to say he was any less impressive ... You see, for a start, as we read back on page 63, Clever Hans was a horse.

In the early 1900s, Hans drew worldwide attention and fame from his Berlin stable. Here, his trainer would demonstrate this ungulate prodigy's amazing ability to not simply count, but solve maths problems by tapping his hooves to indicate a number or correct option among multiple choices.

However, Clever Hans was a fraud.

But before you get on Twitter and cry 'outrage!', bear in mind the horse died well over 100 years ago. Oh, and there's something else you should know: Clever Hans's trainer wasn't in on the trick.

Researchers would later discover that while Clever Hans didn't actually have mathematical skills, the horse *did* demonstrate very impressive observational skills.

It turns out Clever Hans was unable to answer questions that his questioners also couldn't answer because he was actually reading minute facial and body cues to determine the correct responses.

Clever Hans, indeed.

'Without mathematics, there's nothing you can do. Everything around you is mathematics. Everything around you is numbers.'

—Shakuntala Devi

Venus flytraps can count

Incredibly, researchers have observed Venus flytrap plants 'counting' the number of touches made by prey within their 'jaws' to tailor a predatory response.

1 touch =
Pay attention, but don't respond yet.

2 touches =
Probably food — trap closes.

3+ touches =
Start digesting!

The human calculator

Shakuntala Devi was an Indian writer and 'mental calculator'. Intrigued? Well, prepare to be astonished.

To say she was good at mentally calculating the answer to problems is like saying LeBron James can play basketball.

Devi earned a reputation which led to her touring the world demonstrating her talent throughout the mid-20th century.

In 1977, at the Southern Methodist University in Texas, Devi managed to calculate the 23rd root of a 201-digit number *in her head*. Her answer, 546,372,891, took 50 seconds. It was then confirmed by the US Bureau of Standards on a UNIVAC 1101 computer … for which a specific program had to be written to crunch the number.

In 1988, she put her skills before Berkeley professor of psychology, Arthur Jensen. He asked her to perform several calculations including, famously, the cube root of 61,629,875 and the 7th root of 170,859,375. Devi managed to provide the solutions (395 and 15, respectively) before Professor Jensen had even finished noting the numbers down in his notebook.

As you may suspect, Devi made her way into the *Guinness Book of World Records* — in the early 1980s. She did so by correctly demonstrating the multiplication of two numbers, picked at random by the Computer Department of the Imperial College London, in 28 seconds.

Oh, what were the numbers? 18,947,668,177,995,426, 462,773,730, in answer to the question, 7,686,369,774,870 × 2,465,099,745,779.

If you're interested, as I'm pretty certain you are, in Devi's calculations, track down a copy of her book The Joy of Numbers.

Score!

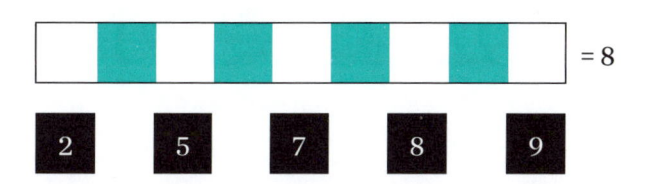

▢ ▨ ▢ ▨ ▢ ▨ = 8

[2] [5] [7] [8] [9]

▢ ▨ ▢ ▨ ▢ ▨ = 8

[3] [6] [7] [8] [8]

▢ ▨ ▢ ▨ ▢ ▨ = 8

[1] [4] [6] [7] [8]

Reach the goal number by creating an equation using the provided numbers and your own choice of operations.

Numbers are placed in white squares, while operations (+, −, ×, ÷) are placed in green squares.

Order of operations matters here, per usual, and you can't use brackets!

Read the numbers off in order for your final score.

What is the equation that gives the highest score?

(No Model.)

S. WHEELER.
WRAPPING OR TOILET PAPER ROLL.

No. 459,516. Patented Sept. 15, 1891.

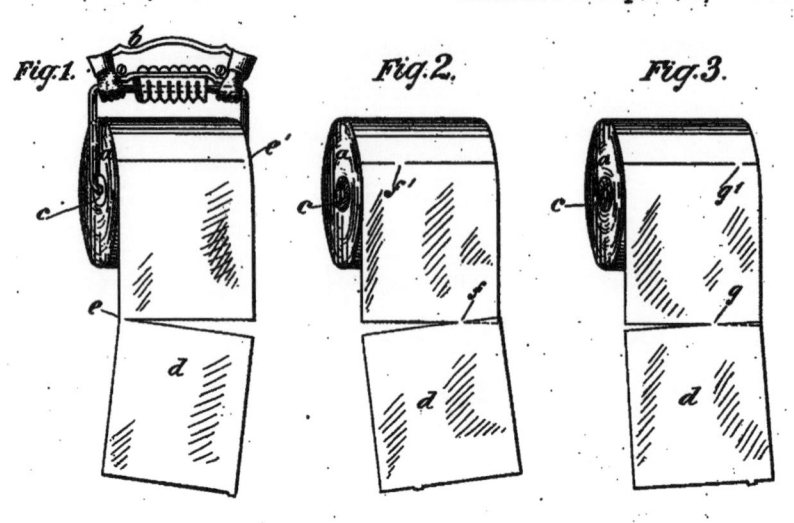

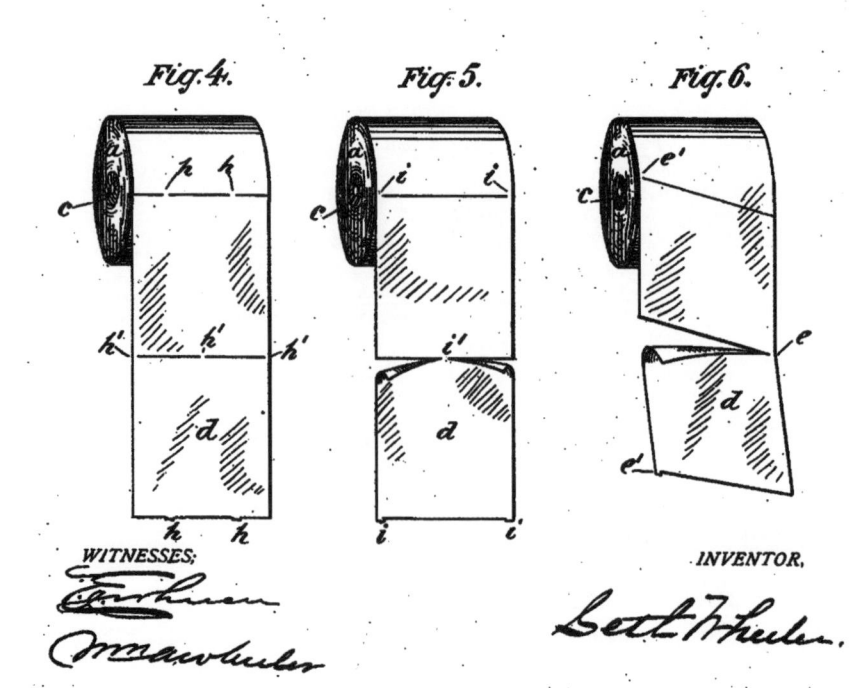

Fig. 1.   Fig. 2.   Fig. 3.

Fig. 4.   Fig. 5.   Fig. 6.

WITNESSES: INVENTOR.

Case closed?

The original 1891 patent for a toilet paper roll clearly shows that the correct way to insert it is with the paper *over* the roll.

Its inventor, American businessman Seth Wheeler, had been frustrated by the 'many devices designed to prevent waste' of paper, 'namely, in the construction of holders for the rolls provided with means to prevent free unwinding of the roll and cause the sheets to separate singly at their connecting points'.

'My improved roll may be used on the simplest holders,' Wheeler proclaimed and, over a century into the future, millions of internet users would endlessly debate the 'over' versus 'under' placement of the roll. Not to mention the 'fold' versus 'scrunch' debate.

According to the venerable* Toiletpaperworld encyclopedia, 40% of us are folders, 40% are 'wadders' and 20% are wrappers. I've got to be honest, I'm not going to think too closely on the subtle distinctions there.

Nevertheless, closer to home, the Toilet Paper Man's website (he specialises in selling ... you guessed it ... online) reckons that 'folders' use 2–3 sheets per use, whereas 'scrunchers' 'typically' use 4–7, making the former significantly more efficient.

Is that data peer-reviewed? Look, at the very least it'll give you something to ponder as you go about your business next time. Why not conduct your own research? A rear-reviewed study (sorry)? By all means report back the data for you and your family on Twitter and add to this, ah, important debate**.

But one thing's for certain, we all owe Seth Wheeler and his 'wrapping or toilet paper roll' patent of 1891 a debt of gratitude.

* Possibly venerable?

** Joking aside, by some estimates, a single tree produces about 3000 rolls of toilet paper. Since about 83 million rolls are produced every day, global toilet paper production consumes over 27,000 trees ... daily.

So anything we can do to reduce the waste can't be a bad thing.

A letter of numbers

Here are a few quickies for you to ponder over your morning coffee.

Or tea, for that matter.

- Which number is spelled with its letters arranged in alphabetical order?
- Which number is spelled with its letters arranged in *reverse* alphabetical order?
- Only one number from 0 to 1000 includes the letter 'a', excluding the word 'and'. What is it?
- Only one number in the English language is spelled with the same number of letters as the number itself. What is it?
- Which anagram of 'twelve plus one' also equals the same answer, thirteen?

Your time starts ... *now*!
    Answers at the back, as usual.

What do you meme?

Here's another one which is, as they say 'DRIVING PEOPLE EVERYWHERE CRAZY!'

I came across it when the awesome Cliff Pickover (@pickover, whose *The Math Book* is geek GOLD) tweeted it out.

According to the original tweet, 'This problem can be solved by pre-school children in 5 to 10 minutes, by programmers in an hour and by people with higher education … well, check it yourself!'

No mention of the usual claim that solving it in under X number of seconds means you're a genius, but it's a curious little puzzle to tickle your brain.

Your task, should you choose to accept it, is to solve it.

8809 = 6	5555 = 0
7111 = 0	8193 = 3
2172 = 0	8096 = 5
6666 = 4	1012 = 1
1111 = 0	7777 = 0
3213 = 0	9999 = 4
7662 = 2	7756 = 1
9313 = 1	6855 = 3
0000 = 4	9881 = 5
2222 = 0	5531 = 0
3333 = 0	2581 = ?

d'amazing Leo

It goes without saying that entire books (and more) have been written about Leonardo da Vinci, one of history's most intriguing characters, so I'm not even going to attempt to summarise the man's genius here.

I mean, he excelled in literally dozens of fields, including painting, architecture, astronomy, engineering, botany and of course, mathematics ... The Florentine polymath epitomised the term 'Renaissance Man', painted the *Mona Lisa* and *The Last Supper*, designed the first flying machines, calculators and tanks and ...

See what I mean? Just too much genius to cover in this book. So let's drill down to something I find truly beautiful and truly relevant to our journey through *Numberland*. Yes, my shape-shifting friends, I want to talk polyhedra. Keen readers will recall a polyhedron is a solid shape in 3 dimensions with flat polygonal faces, straight edges and sharp corners (also known as vertices). The word comes from the Greek — 'poly' meaning 'many' and 'hedron' meaning 'face'.

Anyway, Leonardo loved geometry — who doesn't — but he only got into it in a big way when he was in his 40s. Scholars say some of his best geometric work can be found in a book from 1509 called *The Divine Proportion* by Luca Pacioli. The book features around 60 gorgeous hand-drawn illustrations of various polyhedra (including our pals the Platonic solids and the 6 Archimedean solids) and could very well be first illustrations of these solids ever!

There's some debate as to whether Leonardo actually built the wooden models on which the illustrations were probably based, but that's small potatoes IMHO.

In celebration of Leonardo's achievement, I think it's worthwhile taking a closer look at his gorgeous old-school representations.

These polyhedra have been scanned and 'cleaned up' a bit for printing by Dr Robert Fathauer. When I asked Dr Robert about reproducing his homages to Leonardo in this book, his reply was swift: 'That's fine — just don't make it sound like I'm so arrogant as to think I can improve on the work of Leonardo da Vinci!'

It's certainly a tall order to try to compete with the Old Masters, but when it comes to beautiful representations of maths in art, Dr Robert's certainly no slouch. Head over to page 364 to find out more!

De scales

I love a good scale.

I don't mean a progression of musical notes — although they also rock ... *rock* ... geddit?

Nor do I mean a bathroom scale — although I did lose 17 kilos in 3 months earlier this year. My secret? Well, it turns out diet is important! Science in action, folks.

What I mean by 'scale' in this instance is a list of numbers that rank something. Like, say, the Richter scale* which measures earthquake intensity. Or (from my last book) the Beaufort Scale of Wind Speeds. Or from a couple of years back, the Bristol Stool Chart, by which you measure the health of your poo — or more particularly your own health given where on the scale your deposit sits.

So let me present the following 3 scales you:

1. Never
2. Vaguely
3. Reasonably
4. Definitely

... knew existed!

The Rio Scale of Alien Encounters

First developed in 2001, the Rio Scale is used by astronomers seeking signs of extraterrestrial life. It helps to communicate to the public 'how excited' they should be about the phenomenon observed.

Though not yet ratified by the International Academy of Astronautics Permanent Committee of SETI (Search for Extraterrestrial Intelligence), it had indeed been submitted at the time of writing. Stay tuned.

---

* The Richter scale measures the shaking intensity associated with an earthquake, quantified by the amplitude of vibrations on a device called a seismograph.

A magnitude 5.0 quake has an amplitude 10 times larger than a magnitude 4.0 quake.

The Richter scale has largely been superseded by the Moment Magnitude Scale (MMS) due to the inherent difficulties of measuring large quakes. It's a different logarithmic formula which correlates to measurements on the Richter scale.

10. Extraordinary
9. Outstanding
8. Far-reaching
7. High
6. Noteworthy
5. Intermediate
4. Moderate
3. Minor
2. Low
1. Insignificant
0. None

Mohs' Scale of Mineral Hardness

This one was named after Friedrich Mohs, which seems fair since he invented it.

Mohs' Scale is based on the ability of one mineral to scratch another. There are 10 minerals included on it, ranging from the super-soft talc, through gypsum, and calcite, onto fluorite then apatite, and my personal favourite — feldspar — before encountering some serious hardness with the double zeds of quartz and topaz. In second place — so close to number one! — you've got quite a conundrum if you're trying to beat corundum. While at number one, the Mohs' bada$$ mineral on the Mohs ... come on down, diamond!

1. Talc
2. Gypsum
3. Calcite
4. Fluorite
5. Apatite
6. Feldspar
7. Quartz

8. Topaz
9. Corundum (for example, ruby or sapphire)
10. Diamond

The Okta Scale

This nifty little scale quantifies the amount of cloud cover over a given location.

Estimations for the sky conditions are based on how many 8ths of the sky are covered by cloud, ranging from completely clear skies (0 oktas) to completely overcast (8 oktas).

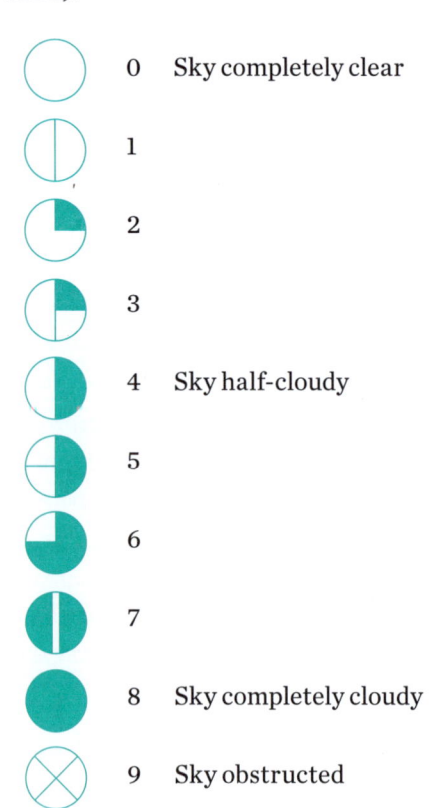

Symbol	Value	Description
○	0	Sky completely clear
◐	1	
◕	2	
◕	3	
◐	4	Sky half-cloudy
◑	5	
◕	6	
◕	7	
●	8	Sky completely cloudy
⊗	9	Sky obstructed

'While physics and mathematics may tell us how the Universe began, they are not much use in predicting human behaviour because there are far too many equations to solve. I'm no better than anyone else at understanding what makes people tick.'

—Stephen Hawking

Off to see the Queen

Puzzle number 125 in Smullyan's wonderful book (check out page 159) sees the reader off to visit the truth-telling King.

However, given we're in *Numberland*, it seems more appropriate we head off to see the Queen. Because here, the Queen *always* tells the truth.

She tells us that nearby there is a crossroad. One route will lead to untold treasure. The other to something far more unpleasant — say, a village which only plays Ed Sheeran on repeat.

Of the 3 guards guarding the crossroad, one always tells the truth. One always lies. The third lies at random. The 3 of them know how the others behave.

What *two* questions, each directed to 1 of the 3, will lead us to the untold treasure?

The most dangerous number ...

According to the sharp-eyed journos at *Rolling Stone* magazine, I am sad to say we have a contender for the 'most dangerous' number.

You may be thinking it's '666' (if you're mythologically inclined). Or '000' or '911' (if you're doing something naughty). You might even detour to the 1937 comedy film *Dangerous Number* starring Robert Young and Cora Witherspoon (no relation to Reece) ... if you're feeling tangential.

But no, the most dangerous number is, in fact ... 415. Why? Well, in May 2019, atop Hawaii's Mauna Loa volcano, scientists recorded carbon dioxide levels of 415 parts per million. That's not a particularly large number compared to say the distance between the Earth and the edge of the observable Universe (around 46 billion light years), but it *is* the highest recording of carbon dioxide levels since human beings have lived on Earth.

The last time they were this high was back in the Pliocene epoch, roughly 3 million years ago.

One other 'dangerous' number? 33,565,835,380. That's the approximate weight, in tonnes, of carbon dioxide we dumped into our precious atmosphere in 2018 as a result of burning fossil fuels.

ADAM SPENCER

According to the elves @qikipedia, people who have regular sex feel less stressed when doing mental arithmetic

NUMBER
LAND

I guess they must've done the math ...

~~Four~~ Three score

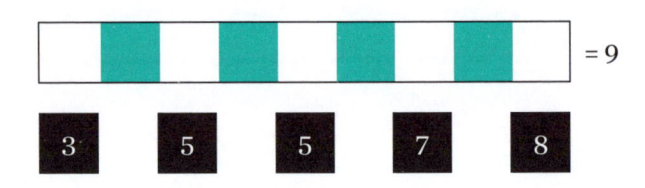

= 9

| 3 | 5 | 5 | 7 | 8 |

= 9

| 2 | 3 | 4 | 7 | 9 |

Reach the goal number by creating an equation using the provided numbers and your own choice of operations.

Numbers are placed in white squares, while operations (+, –, ×, ÷) are placed in green squares.

Order of operations matters here, per usual, and you can't use brackets!

Read the numbers off in order for your final score.

What is the equation that gives the highest score?

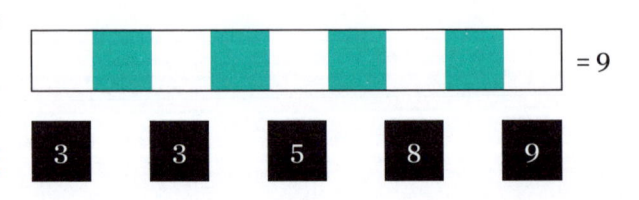

= 9

| 3 | 3 | 5 | 8 | 9 |

318

Going, going, gone ... again!

Another unenviable statistic that cannot be repeated enough is Australia's woeful record in making our furry friends extinct.

And I'm not talking about bearded baristas here but rather our native fauna.

When it comes to wipe-outs, we are 'top' of the charts: as at the time of writing, some 30 of our mammals are gone, accounting for 35% of the global extinct population. Since 1788, as a result of hunting, deforestation (reduction of habitat), introduction of exotic predators, as well as changes in the environment, we've lost or badly endangered the:

1. Boodie, burrowing bettong
2. Brush-tailed bettong
3. Desert rat-kangaroo
4. Pig-footed bandicoot
5. White-footed rabbit-rat
6. Central hare-wallaby
7. Rufous hare-wallaby
8. Eastern hare-wallaby
9. Banded hare-wallaby
10. Tammar wallaby

11. Toolache wallaby
12. Lesser bilby
13. Bramble Cay melomys
14. Long-tailed hopping mouse
15. Big-eared hopping mouse
16. Darling Downs hopping mouse
17. Crescent nail-tail wallaby
18. Western barred bandicoot
19. Desert bandicoot
20. Broad-faced potoroo
21. Blue-grey mouse
22. Gould's mouse
23. Dusky flying fox
24. Maclear's rat
25. Bulldog rat
26. Tasmanian tiger
27. Christmas Island shrew
28. Lesser stick-nest rat
29. Lord Howe long-eared bat
30. Christmas Island pipistrelle

Nice work, us.

And these are just the animals we know of. There are likely many more. Reminding ourselves about these creatures, how they lived and why they die out helps scientists develop strategies to ensure no more go the same way. It turns out, one of the key determinants of survival is weight.

According to Professor Richard Kingsford from the University of New South Wales, the most precarious 'weight range' is between 35 grams (about the size of a large mouse) and 5.5 kilograms (about the size of a wallaby). Creatures in this range are 'bite-sized' and particularly vulnerable to human beings, cats and foxes (all introduced species). While most humans know better, cats and foxes don't, and are superb predators capable of decimating entire populations of native fauna in no time at all.

One solution is the creation of more fenced-off, inland natural areas called 'exclosures' — sort of like huge, wild zoos — to keep the feral predators out and the native fauna safe. Another is situating the exclosures in the vicinity of dingoes, who eat cats and foxes.

Scientists aren't sure we can completely eradicate all the feral animals (it's guesstimated there are more than 20 million cats and 7.5 million foxes running wild) but there's a growing body of evidence that suggests some of our native animals (like bilbies) can tolerate low numbers of exotic predators and actually learn how to outsmart them.

Once again, it's all down to the numbers. Twenty-seven and a half million feral animals is clearly too many, but if we can reduce that number — and increase the number of exclosures — we may well slow the extinction rate. I think that's better than force-feeding bilbies so they outgrow that precarious 'weight range'!

☐ * ☐ * ☐ * ☐ * ☐ * ☐

Another Tanton terrifier

In this next fiendish puzzle from our marvellous mathematician James Tanton, each * signifies either a < or >.

For each pattern of choices of <s and >s, can you fill the boxes with the numbers 1, 2, 3, 4, 5, 6 to respect those inequalities? Which pattern of <s and >s gives the most solutions? And how many?

Answers, as always, are at the back of the book.

HINT: There are 32 possible solutions.

'Read Euler, read Euler, he is the master of us all.'

—Pierre-Simon Laplace

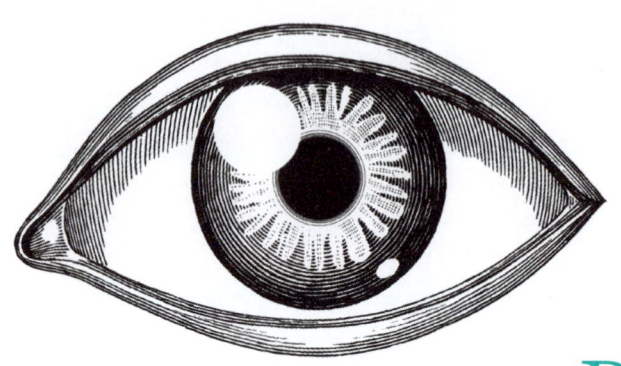

Blink and you'll miss it

Have a guess how many times you blink in a minute.

Fifty? A hundred? It's more like 20 times, as it turns out. That equates to 1200 blinks per hour and 28,800 per day.

Now, it probably won't come as a surprise that based on an average night's sleep of 8 hours, you'll end up spending 25 years of your life slumbering — assuming you live to the age of 75, that is.

But what might be a surprise is the fact that we spend an additional 10% of our waking hours with our eyes closed (while we blink)!

Happily, the average human blink only lasts for about 400 milliseconds. So yes, you are missing out on stuff.

But not all that much.

Belgian waffles

If you or I break a toaster, or crack the screen on our phone, it's irritating. But you wouldn't call it catastrophic.

Spare a thought for the poor Belgian Air Force maintenance men who, like the hapless Red Bull-loving co-pilot you'll meet on page 347, cost their government (and the Belgian taxpayer), an aircraft hangar's worth of money.

In October 2018, at the Florennes Air Base in the southwest of the country, a couple of guys were working on an F-16AM fighter jet when they accidentally fired the plane's 20-millimetre, 6-barrelled Gatling gun (a type of rotary cannon). The cannon rounds hit another F-16AM fighter which was parked nearby, fully fuelled and ready for flight.

The fighter exploded into flame and, in a matter of minutes, was completely destroyed. Even worse, two fellow Belgian Air Force personnel were treated for injuries, while the maintenance guys who accidentally fired the gun suffered hearing damage.

Even though many air forces around the world are phasing out their F-16 fleet, they're still incredibly sophisticated — and expensive — machines. When they were brand new, they cost in the vicinity of US$15–20 million.

That's a *lot* of waffles.

Tile high club

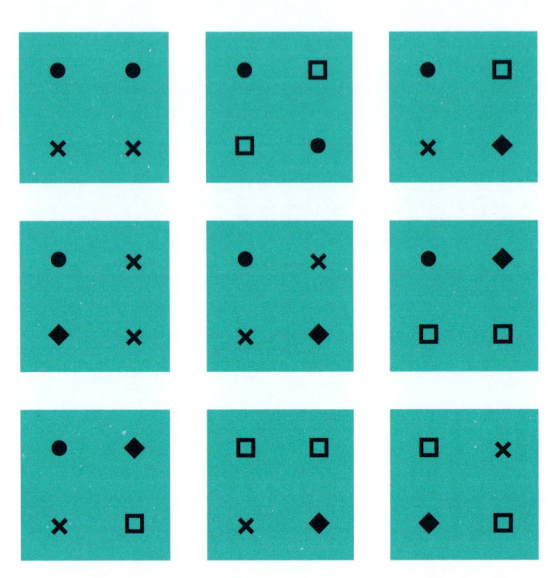

Congratulations! You've reached our final tiling puzzles.

Hopefully as I've slowly reduced the clues, you've improved your handy-person skills to the point where you can do these two examples completely hint-free.

Remember to place the 9 tiles into a 3 × 3 grid so that adjacent tiles show the same symbol along their touching edges.

The tiles may need to be rotated, but they are never reflected.

Grab the grout and tile on!

As always, you can head over to my website to download a cut-uppable copy of this puzzle.

If you've enjoyed these tiling terrors, you'll find many more in my previous book, *Top 100*.

Point your browser to adamspencer.com.au

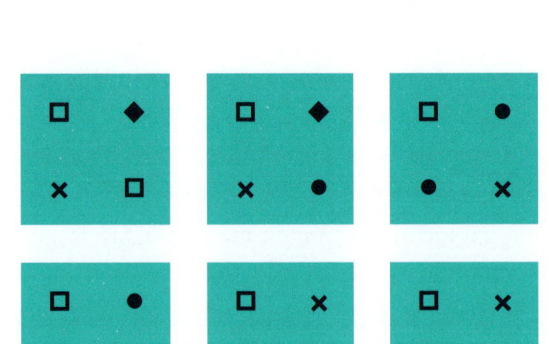

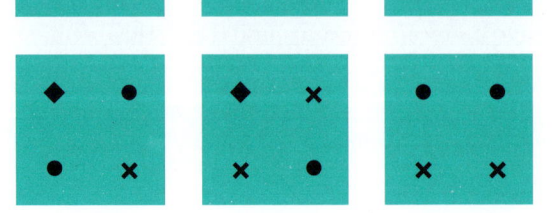

326

Place your bets!

G ambling is generally a pretty bad idea.

Okay, had to put that out there straight up.

But there are branches of mathematics that incorporate cool aspects of games of chance. Take the St Petersburg Paradox, for example.

Let's say you're at a casino (only if you're over 18, of course). The generous casino boss invites you to play a game where the pot of money starts at $1 and on each turn a coin is tossed. If the coin comes up heads, then the pot is doubled, if it comes up tails, then you win whatever's in the pot.

The question is: how much would you pay to play this game?

Let's break it down. Fifty per cent of the time tails come up on the first toss and so you pocket $1. But the other 50% of the time heads come up, the pot doubles to $2, and you get to toss the coin again.

On your second toss, 50% of the time you get tails and win the $2 and 50% of the time you get heads and the pot doubles to $4. So you get to toss again.

A pattern emerges: 50% of the time you win $1, 25% of the time you win $2, one eighth of the time you win $4, and so on. The amount you win gets bigger and bigger, but the chances of winning that amount get smaller and smaller.

You should expect to win:

$$(1/2 \times \$1) + (1/4 \times \$2) + (1/8 \times \$4) + (1/16 \times \$8) \dots$$

There are an infinite number of terms in this equation.

Which is:

50c + 50c + 50c ...

Which is the same as an infinite number of 50 cents, which is infinity.

So you might very well expect to pay an infinite amount of money to play the game because you will win an infinite amount of money!

Or would you? Herein lies the 'paradox' part of the St Petersburg Paradox.

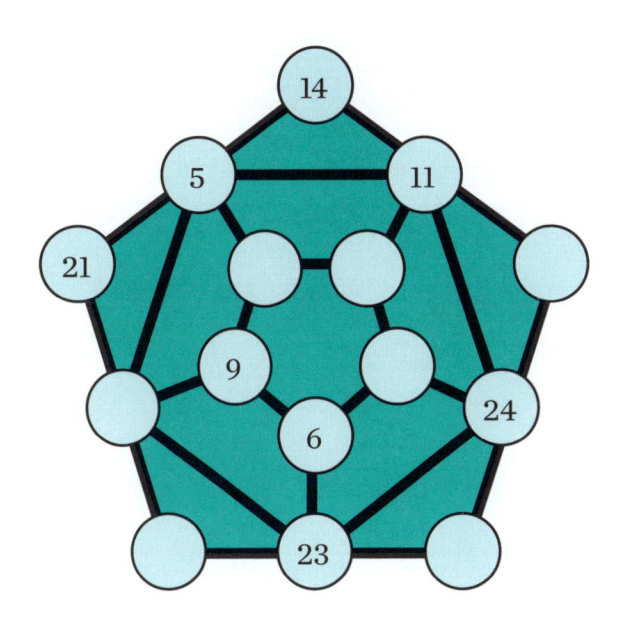

Questacontent 5

Q uestacon's dastardly puzzlers provide this pentagonal paradox for your perusal ...

The Pentagon in Washington is so named because of — you guessed it — its shape. We know[*]:

- The corridors divide the building into 8 different pentagons.
- To monitor movement throughout the building, there are 15 security control points.
- These security control points are positioned so that, for each pentagon, the total number of guards is always 80.
- First, locate the 8 pentagons. Some of the guards, as you can see, are already at their control points.
- Find how many guards are positioned at each of the remaining control points.

Off you go!

[*] Actually, to be honest, I get the distinct impression that this stuff is kept a bit secret by the Pentagon. Just sayin'.

HINT: The missing number of guards at each station are 8, 12, 15, 17, 22, 25 and 28.

329

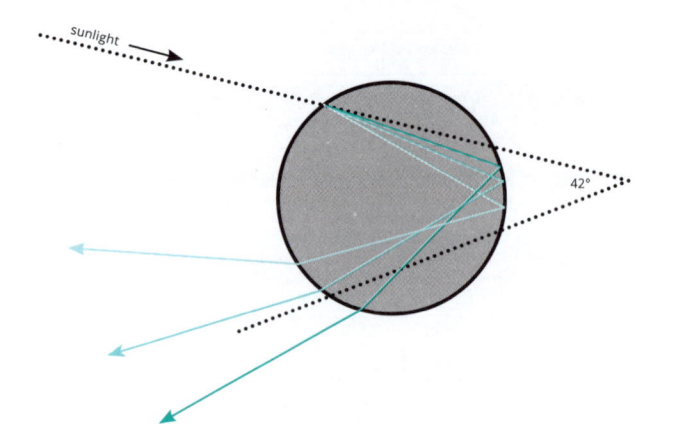

Raincircles

The next time you see a rainbow, take a good look at it and try to determine its shape.

Most people would say rainbows are semi-circular. In fact, they're round. To understand why this is so, we need to first get our heads around two important concepts when it comes to light: colour and refraction.

Even though sunlight might appear just the one 'colour', it's actually made up of lots of different colours mixed together in 'waves' of various lengths, from the longest wave (red), through orange, yellow, green, cyan, blue and indigo to the shortest wave (violet).

Now, when sunlight goes through water, it changes direction. The light waves actually 'bend' and the amount of bend is determined by the wave's colour. Short waves bend (or change direction) more than long waves. So indigo bends more than orange.

This 'bending' process is called refraction and a rainbow occurs when light waves travel through water (for example, rain or mist) and break apart into the various lengths, revealing all the colours (of the rainbow!).

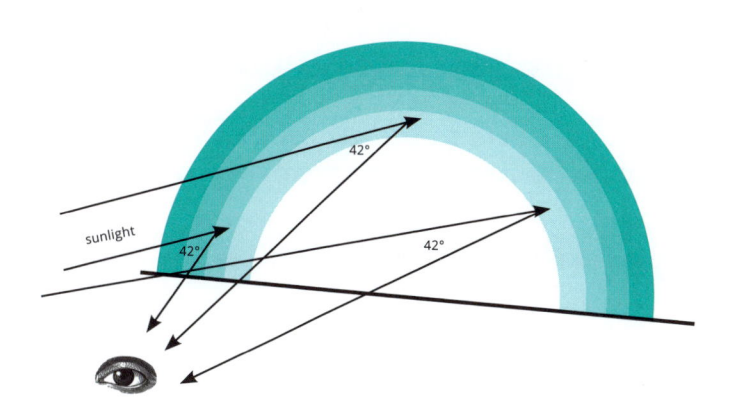

But Adam, you told me rainbows were round not semi-circular. Correct.

Take a look at my extremely rare, all-green rainbow diagram above (hey, you try drawing a rainbow diagram with only green and black ink at your disposal!)

The other thing that happens when sunlight travels through water is that some of that light 'bounces back' — or is reflected. Sunlight bounces back most strongly at an angle of 42° so if you're on the ground and the sunlight reflects at that precise angle into your eyes, you'll see a rainbow. You can only see half the shape, though, because the Earth is blocking the other half. At altitude, when you're in a plane, for example, you may be lucky enough to see the entire thing. And it's circular.

Don't know about you, but 'raincircle' might take a bit of getting used to.

Vending Vendetta

Nope, that's not the title of a new James Patterson novel.

Turns out there's a predator in the US far more dangerous than the shark. You're probably thinking grisly bear. Or angry moose. Or any number of crazy human beings wielding weapons. They can all certainly be dangerous.

But there's another malevolent force lurking much closer to home. Like, in your school or office.

I'm here to tell you the humble vending machine is killing and maiming 4 to 6 times as many people as sharks in the US of A! And that's no urban myth. Each year, sharks attack around 25 people. On the other hand, in 2018, there were over 1700 reported injuries from vending machines.

Between 2002 and 2015, four people were actually killed by the mechanical monsters, making them the 10th out of 15 deadliest objects in American schools and offices. (If you were wondering, monkey bars and other playground equipment topped the list as the most dangerous. Scissors came next.)

Vending machines prefer older victims — 30% of all injuries happened to people over 60 years of age. Men also copped it worse at around 55%. Most injuries were to the head, the hands, the upper body, the face and, when a vending machine falls over, the whole body.

On a happier note, the majority of VMIs (vending machine injuries) didn't require a visit to the hospital. Hmm, bring back the trusty tuck shop, I say.

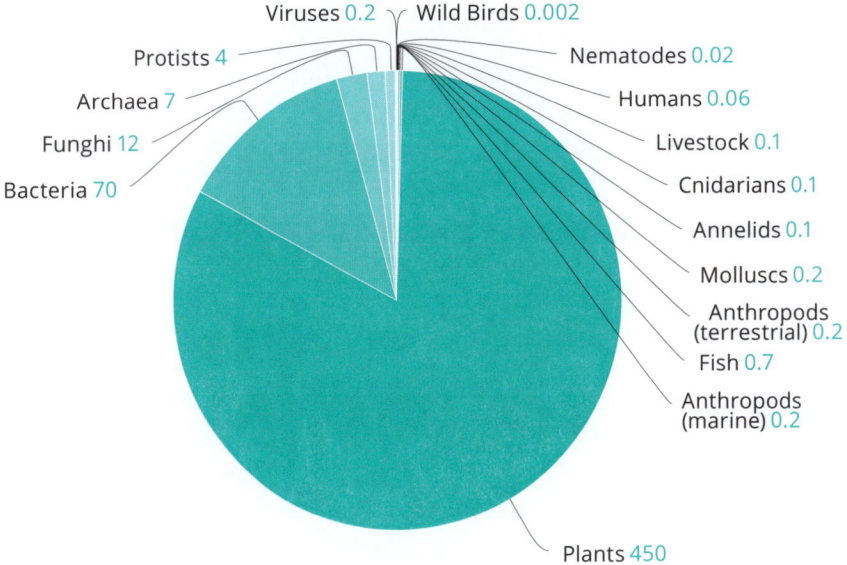

Pie chart labels:
- Viruses 0.2
- Wild Birds 0.002
- Protists 4
- Nematodes 0.02
- Archaea 7
- Humans 0.06
- Funghi 12
- Livestock 0.1
- Bacteria 70
- Cnidarians 0.1
- Annelids 0.1
- Molluscs 0.2
- Anthropods (terrestrial) 0.2
- Fish 0.7
- Anthropods (marine) 0.2
- Plants 450

Heavyweight champs

The brilliant boffins at the National Academy of Sciences in Washington DC have worked out that the carbon weight of all life on Earth is around 550 gigatonnes. A gigatonne is 1 billion tonnes (or 10^9 tonnes or 1 trillion kilograms).

Of that 550 gigatonnes, we human beings make up a tiny percentage, only around 0.06 gigatonnes (which is 0.01%). There's considerably less of 'us' as there is of 'them'! And by 'them', I mean fish (0.7 gigatonnes); crabs and lobsters (1 gigatonne); mushrooms (12 gigatonnes); and bacteria (70 gigatonnes).

But the undisputed heavyweight champions of the world are … plants. Yep, at 450 gigatonnes, our natural flora accounts for the biggest proportion of the Earth's biomass.

Which doesn't mean we can indiscriminately destroy them, mind you!

All you can eat

Bodybuilder turned Ironman triathlete Jaroslav Bobrowski sure has an appetite.

To be expected, I guess. He's super fit and follows a special diet in which he fasts for 20-hour periods. When he does sit down for dinner, he understandably has to make up for lost time.

Back in September 2018, he rocked up to his local sushi train in Lanshut, Germany, to enjoy their 15.90 euro all-you-can-eat meal deal. *Sehr gut*, baby! Ol' Jaroslav did a few squats and crunches, took a deep breath, and proceeded to demolish almost *100 plates of food*. He washed it down with tea as his strict training regime doesn't permit alcohol.

Now, an average serving of sushi contains around 40 to 50 calories, meaning Jaroslav ate around 4000 calories in just one sitting. The 100 plates of sushi weighed around 8 kilograms. That's an extremely unhealthy amount of food to consume in one sitting, but mind-boggling considering Jaroslave is just 173 centimetres tall and weighs 78 kilograms.

Well, you don't need me to tell you the restaurant owner Tan Le was not impressed. He banned Jaroslav from ever coming back for seconds.

Snap! Crackle! Pop!

Ever wondered where the famous cereal mascots' names truly came from?

Yep, they do sort of sound like the noises we make when eating breakfast, but I prefer the idea that they're based on physics.

You see, 'snap', 'crackle' and 'pop' are the terms physicists use for certain time derivatives of position. The 1st derivative is velocity, the 2nd is acceleration, and the 3rd is jerk. The 4th is snap, while the 5th and 6th are sometimes called crackle and pop.

I get why Kellogg's went with the 4th, 5th and 6th derivatives. *Velocity! Acceleration! Jerk!* doesn't have quite the same breakfasty ring.

Ferrets are nocturnal and sleep for up to 20 hours during the day

The popular perception of the ferret is of a busy creature constantly darting around — literally ferreting for food, shelter, friends — or mischief.

In fact, animals in the *Mustela* genus are nocturnal and spend the daytime sleeping for up to 20 hours.

They're so good at deep slumber that many an owner has mistakenly thought their pet was in a coma ... or worse, dead.

Nope. Just recovering from all that ... ferreting.

Get your groove on ... while you still can

The jury's back and the verdict seems to be a lot of us stop listening to new music around the time we turn 30.

Research conducted in the UK revealed that our peak age for discovering new tunes is around 24. After that, due to a combination of factors including too much choice, nostalgia, work and family commitments, we tend to stop grazing widely and go back to what we know (and love). Again and again.

Economist Seth Stephens-Davidowitz analysed a huge amount of Spotify data and found that if you were in your teens when a song was first released, it'll be the most popular among your age group 10 years later. Our favourite songs release a flood of feel-good chemicals like dopamine, serotonin and oxytocin in our brains and while this can happen at any time, when it happens when we're teenagers — and super hormonal and sensitive — chances are whatever ditty was responsible will be with you for life.

> Come on, Eileen, oh I swear (what he means)
> At this moment, you mean everything ...

Take your pica

I don't need to tell the typographically inclined among you that a 'pica' is 1/6 of an inch and there are 12 'points' to a pica.*

But do you know where the terms originate?

Back in 1737, the French typographer Pierre Fournier came up with a measurement system for text. He used a 12-point unit called a 'cicero', equivalent to 0.1648 inches. A 'point' therefore was a unit of length equal to 0.0137 inches.

But in 1770, the printer François-Ambroise Didot converted Fournier's system so it aligned with the legal French foot at the time, creating a larger 0.1776-inch 'pica', with 12 points each measuring 0.0148 inches.

Towards the end of the 18th century, France went metric, but Didot's system was influential and is still widely used in Europe to this day.

Oh, and 'pica' means 'magpie' in Latin. In Anglo-Latin it evolved to also mean a church book of rules, either because the book was a collection of miscellanea (much like magpies collect objects), or because both were black and white in appearance. Take your pica, I guess … ? Sorry.

* In case you were wondering, this book is set mostly in the Chronicle font, designed for the New Times newspaper chain by Jonathan Hoefler and Tobias Frere-Jones back in 2002.

Ol' 55

Here's another brain buster for you.

Make a sequence by repeatedly subtracting the biggest square you can until you hit zero.

For example, let's consider 55. The nearest square less than 55 is $7^2 = 49$. Subtracting 49 from 55 we get 6, and continuing this process will give you the following 5-term sequence:

$$55, 6, 2, 1, 0$$

Tell me, what is the *smallest* positive integer that will give you an *8-term* sequence?

Answer at the back of the book.

'The power of mathematics is often to change one thing into another, to change geometry into language.'

—Marcus du Sautoy

Unlucky number 13?

Anyone who's read any of my books will know that I'm not superstitious.

I really don't go in for loopy mythological theories, folkloric ideas and untested notions based on hearsay, hunches or guesses. But if you *are* of a superstitious bent — or if you happen to be triskaidekaphobic — then here are 13 reasons why you might be wary of the number ... 13.

1. 1979's US Super Bowl XIII was an absolute shocker for the bookies when the Dallas Cowboys lost to the Pittsburgh Steelers.
2. Staying with American footy, one of the best ever players, quarterback Dan Marino, never won a Super Bowl. Why? He wore the number 13 jersey ...
3. The ancient Zoroastrians predicted there was gonna be chaos in the 13th millennium.
4. 12 is the sum of the divisors 6 — a perfect number. Trying to improve on perfection (twice over!) seems a tough ask; 13 *must* be unlucky.
5. Many businesses prefer to lay low on Friday the 13th. It's estimated this costs the global economy billions each year.
6. Some people believe having 13 letters in your name is bad. Hmm, Charles Manson, Jack the Ripper, Jeffrey Dahmer, Theodore Bundy, and Albert De Salvo, people? But what about that Nazi leader. Well, Hitler's baptismal name was actually *Adolfus Hitler*.

7. The preferred number of witches for a coven is ... 13.
8. There's a Nordic myth about 12 gods having a dinner party at Valhalla which is crashed by a 13th guest, Loki, the god of mischief. Loki arranged for the god of joy and gladness to be shot with a mistletoe-tipped arrow and when he died, the world grew dark and sad.
9. The alleged protectors of the Holy Grail, the Knights Templar, were rounded up by Pope Clement V on 13 October 1307.
10. Traditionally, there were 13 steps leading up to the gallows.
11. The Bible says that the Jews murmured 13 times against God in the exodus from Egypt, that the 13th psalm is about wickedness and corruption, and that the circumcision of Israel occurred in the 13th year.
12. Still on themes Biblical, many people believe the Last Supper occurred on the 13th.
13. And while we're there, let's end on the belief that there were 13 people at said Last Supper. The 13th and last person to take a seat? Judas.

There you go. Of course there's a technical term you can use to describe such phenomena as these: coincidences.

I can also think of a term you can use to describe people who believe this kind of stuff.

And no, I'm not thinking of 'superstitious' ...

* Just so we don't end on an 'unlucky' number of theories, I'll also tell you that Dublin's 1875 Chamber Street fire released thousands of gallons of whiskey into the streets. Thankfully no one died in the fire ... but 13 people did perish due to alcohol poisoning! Big shout-out to the QI Elves for that one.

Wheel o' Ferris

Unless you're petrified of heights, at some point in your life you've probably laughed, cried, smooched or prayed while sitting in a cabin atop an amusement park Ferris wheel.

There's just something whimsical about rotating slowly high above the ground while taking in the views.

But if you're anything like me, Ferris wheels aren't just quaint engineering marvels from a bygone era, they're treasure troves of mathematical intrigue.

Say what? Well, one evening, while at the amusement park, I befriend the Ferris wheel operator who hands me the controls. Excellent. But no, I'm not going to toy with the hapless kids high in the sky. I see opportunity for some safe counting fun.

I look up and note the first cabin, count it and turn its lights off. I then skip one lit cabin, count the next lit one as (2) and turn its lights off. I then skip two further lit cabins, count the next lit cabin as (3), and so on, skipping n lit cabins around the wheel before labelling the $(n+1)$th lit cabin as $(n+1)$ and turning off its lights.

It turns out that if I wrote down the colour of the cabins I labelled as 1, 2, 3, 4 and so on until no cabins were lit, my list would say 'Red, Green, Blue, Red, Green, Blue', and so on.

If, on the ferris wheel, every Red cabin is opposite a Green cabin and every Blue cabin is opposite a Blue cabin, how many cabins does the Ferris wheel have?

Head to the back to check your answer.

*The Ferris wheel gets its name from one of the first engineers to build one, George Washington Gale Ferris, Jr, who constructed an 80-metre-high wheel for the 1893 World's Columbian Exposition in Chicago. GWG Ferris began his career building bridges but soon diversified!

The biggest Ferris wheel in the world is the High Roller in Las Vegas. It stands an impressive 167.6 metres tall, has a diameter of 158.5 metres and twenty-eight 40-person cabins.

That means a total of 1120 passengers can be taking in the views at any one time.

Networking in the Blue Zones

Did you know that the average number of years humans are currently living to is 71.4?

This means there are a lot of people who live far fewer years, but also a lot who are kicking on well beyond the century.

Scientists call the world's regions with the highest longevity 'Blue Zones'. And some of the bluest Blue Zones include the Italian island of Sardinia, Nicoya in Costa Rica, Loma Linda in California, Ikaria in Greece and the Japanese island of Okinawa.

Let's consider Okinawa.

The average life expectancy of Okinawan women is around 90, with many surpassing 100. How do they do it? Well, diet is important and, besides avoiding fatty, sugary and salty processed foods, they eat plenty of low-carb tofu, seaweed, fruits and vegetables, with a bit of pork and seafood every now and then.

They keep fit and active, too. But they don't go to gyms — they practise karate, tai chi, ride bikes, potter around in their vegetable gardens and regularly hit the bathhouses.

Okinawans are also very social and spend a lot of time bonding with friends and family. Most people are supported by what's known as a *moai*, a group of 5 or so buddies who offer emotional, logistic and even financial help.

I reckon they're onto something.

Pu-239

The brilliant American chemist Glenn T. Seaborg accomplished a vast amount during his 86 years on the planet.

Aside from writing over 500 articles and books, he received some 50 honorary doctorates, awards and honours, gave his name to the asteroid 4856 Seaborg, and shared the 1951 Nobel Prize in Chemistry. The element seaborgium is also named after him. But his was not a career without controversy.

On one occasion, while being cross-examined during a US Senate investigation into nuclear energy, he was asked, 'How much do you *really* know about plutonium?'

In his trademark, quiet way, Seaborg answered, 'Sir, I discovered it.' Big statement. But totally true. In February 1941, while working at UCLA, the 29 year old worked out how to isolate plutonium-239 — the primary isotope used for the production of nuclear weapons. He did it by bombarding uranium with deuterons, the nuclei of the hydrogen isotope deuterium. For anyone who's wondered why plutonium was not given the chemical symbol 'Pl', Seaborg chose the symbol 'Pu' because it sounded to him like a kid's description of a really bad smell.

The symbolism proved prophetic. In 1942, Seaborg was drafted into the Manhattan Project, where he led a team of 100 scientists in working out a way of chemically separating fissionable plutonium. And so the die was cast.

Seaborg thought the creation of the atomic bomb was necessary — if the Americans hadn't built it, Nazi Germany might have done so — but he vehemently opposed its use on innocent civilians. With a group of other scientists, he even wrote to President Truman urging him to demonstrate the bomb's power somewhere deserted.

He advocated nuclear disarmament until the day he died, but there's no denying the history books will forever link his name with plutonium's nefarious uses.

Red Bull ... gives you a headache

Keen readers will recall I'm a bit of a fan of certain cold caffeinated beverages. Not naming names or anything.

But you had to feel for the poor US Air Force co-pilot who accidentally spilled a 16 ounce can of Red Bull all over the console of his MC-12W Liberty spy plane back in June 2017. Doing something like this on the ground is bad, but this guy did it while they were in the air!

Thankfully, the pilot quickly shut down the system power and the plane returned safely to base. But the damage was substantial — repairing the console cost the US Air Force (and indirectly the US taxpayer) a cool US$113,675. That equates to over US$7000 of damage for each ounce of spilled fizz, making it quite possibly the most expensive can of Red Bull ever made.

The Air Force came up with a solution, though — a very special flying mug ... costing US$1200 a pop. Why so expensive? Well, everything that goes into an MC-12W cockpit needs to be highly engineered. And although soft drink cans are pretty nifty designs in their own right, they just don't cut the mustard when it comes to US$38 million aircraft.

Dragonflies

Just like human fingerprints, every dragonfly wing is unique, but it turns out that behind the wings' delicate pattern of veins ... there lies maths! Well, a mathematical theory, anyway.

Allow me to explain. Dragonfly wings have two types of veins — primary veins, which are usually fairly long and straight and are in roughly the same place on each individual; and secondary veins, which are smaller and in fractionally different positions on each insect's wings. The combination of the two types of veins divide a dragonfly's wing into lots of little pieces (or sections), a bit like a stained-glass window.

Scientists examined 468 dragonfly wings and calculated the area of each of these sections, labelling them either as circular or elongated. They then recreated the sections by simulating a scaled-down wing to reveal the processes occurring while the dragonfly is growing.

What happened? First off, the primary veins divided the wing into large sections. Second, the scientists randomly selected evenly spaced locations, called 'inhibitory centres', within each wing. They then used a mathematical mechanism called a Voronoi tessellation to select locations for the secondary veins. This sectioned off a region around each inhibitory centre so that every spot inside a section was closer to its inhibitory centre than to any other. Then the wing was able to 'grow', making some sections more elongated than others.

Turns out the tessellation model worked a treat and pretty accurately simulated the growth of real dragonfly wings. Cool, eh? Similar models have been used to examine zebrafish stripes and lizard spots.

NUMBERLAND

Map data (metal bands per million people):

- Iceland: 341
- Norway: 428
- Sweden: 299
- Finland: 630
- Russia: 22
- Estonia: 138
- Latvia: 48
- Lithuania: 54
- Denmark: 154
- Ireland: 23
- Northern Ireland: 78
- UK: 68
- Belarus: 34
- Netherlands: 127
- Belgium: 101
- Luxembourg: 44
- Germany: 122
- Poland: 80
- Ukraine: 17
- France: 69
- Switzerland: 112
- Czech Republic: 120
- Slovakia: 93
- Austria: 124
- Hungary: 105
- Romania: 18
- Moldova: 8
- Slovenia: 120
- Croatia: 90
- Bosnia: 31
- Serbia: 51
- Kosovo: 14
- Bulgaria: 47
- Macedonia: 44
- Albania: 4
- Greece: 162
- Italy: 97
- Malta: 51
- Portugal: 12
- Spain: 65
- Turkey: 6

Totally metal

Here, according to *Encyclopedia Metallum* (May 2016) are the number of metal bands per million people in Europe.

If you've ever endured a Eurovision competition, you may be aware of Scandinavia's astonishing appetite for all things metal.

On the other side of the spectrum (and not shown here) — North Korea, Cambodia, Afghanistan, Yemen and Papua New Guinea all stand at ... 0.

If you're interested in more things metal — and indeed keeping up with the data, point your browser to www.metal-archives.com and ROCK ON!

YOU ARE HERE

349

49 ... er

You're probably thinking of the American footy team from San Francisco, right?*

As good a guess as that may be, can I draw your attention to the sheer beauty of the *number* 49? Each 18 February spare a thought for the smallest number for which both it and its neighbours are squareful. Squareful numbers are those that are divisible by a perfect square greater than 1.

Not only that, Professor William D. Banks from the University of Missouri has proved that every integer in base-10 is the sum of 49 or fewer palindromes. (Palindromes are words, phrases or sequences of numbers that read the same way backwards or forwards.)

And one more 49er, just in case you were still hungry: in January 2016, we discovered what was then our largest known prime number, $2^{74,207,281} - 1$. This 49th Mersenne prime is called M74,207,281 for short and has 22,338,618 digits. (Oh, you'll have to check my previous books for the definition of a 'Mersenne prime'. Can't give you all the clues!)

* The name '49ers' comes from the prospectors who arrived in northern California in the 1849 Gold Rush. The team's original logo featured a mustached miner dressed in plaid pants and a red shirt, jumping in midair with his hat falling off, and firing pistols in each hand: one nearly shooting his foot, and the other forming the word 'Forty-Niners' from its smoke. Since 1962, however, the 49ers logo has been a simple white 'SF' within a red oval.

More cash prizes!

But wait, there's a catch. Isn't there always?

You have to solve another fiendishly difficult problem. No such thing as a free lunch, as they say ...

Mathematicians Persi Diaconis and Ron Graham are offering a whole US$1000 if someone can disprove the following conundrum. Ready? Let's dive on in.

Just say you have a set of points in 3 dimensions and you need to connect all the points together with the shortest possible network. Don't ask me why you need to do this, you just do. Our points in question are the corners of a unit cube and we want to draw enough lines between these points in space so that you can get from one point to any other point by travelling along these lines. If we carefully choose 7 of the cube's edges to be our connecting lines, it's possible to connect all the points with a network of length 7.

If you add a point to the centre of the cube and connect every original point to the centre, you get a network length of $4\sqrt{3} \approx 6.928$. Here, adding an additional point means you can reduce the size of the minimal spanning network by about 1%. You could do better by adding more points. But how many?

The burning question is by *how much* you can reduce the length of the minimum spanning network by adding extra points. It's currently thought the most you can save is about 21.6%. That is, for any set of points, the ratio of the length of the shortest network with extra points to that of the shortest network without extra points is bounded by the following value:

$$\sqrt{\frac{283 - 3\sqrt{21}}{700} + \frac{9\sqrt{11 - \sqrt{21}}\sqrt{2}}{140}}$$

To win US$1000, can you prove (or disprove) this is the ideal ratio?*

*Sorry to say you *won't* find the answer to this one at the back of the book ...

Leech Lattice

No, that chapter heading doesn't refer to the latest techno-metal band melting Spotify.

They're actually a mathematical innovations in the field sphere packing!

Okay, I'm not making it much easier, am I? Keen readers will remember our exploration of the ways in which spheres (such as oranges) can be packed into a multi-dimensional space. This is important not just for farmers wanting to more effectively pack fruit, but also for anyone wanting to transmit data, particularly error-correcting codes (ECCs). These refer to an encoding scheme that transmits messages as binary numbers so that the message can be recovered even if some bits of it are mixed up. They're used in pretty much all cases of message transmission, especially in data storage where ECCs defend against data corruption.

Well in the same way you can arrange circles on a page or tennis balls in space, as we move up through dimensions you can consider how you pack the relevant versions of spheres. In fact, many brilliant minds have spent a long time wondering how to most efficiently pack spheres in 4, 5, 6 and endlessly higher dimensions.

In 2019, even these brilliant minds were dazzled when Maryna Viazovska of the Swiss Federal Institute of Technology showed the densest way to pack equal-sized spheres in 8 and 24-dimensional spaces. What's truly amazing is that Viazovska's now proven that these arrangements also solve an infinite number of *other* problems about the best configuration for points that are trying to avoid one another.

The fact that packing oranges into 8 and 24-dimensional spaces results in solutions to other problems is mind-blowing. Mathematicians call these two configurations 'universally optimal' — basically the gold standard when it comes to proving a theory.

In each dimension higher than 3, when you stack oranges in a classic pyramid structure, the gaps between the oranges grow. When you reach dimension 8, however, you suddenly get room to fit more oranges into the gaps. What you get is a highly symmetric configuration called the E8 lattice. Similarly, when you reach dimension 24, you get a thing called the Leech lattice.

Obviously it's a bit of a headspin to try to visualise things in 8 dimensions, but when you do, the results can be really cool.

The beautiful Higman-Sims graph above gives us an insight into the relationshipo between certain 24-dimensional points inside a Leech lattice.

Look, don't worry. I can't really picture any of this either. But, wow, the Higman-Sims graph would make an awesome neck tatt!*

* I presume I don't have to explain I'm using this as a figure of speech. Please do *not* get a Higman-Sims neck tatt.

ADAM SPENCER

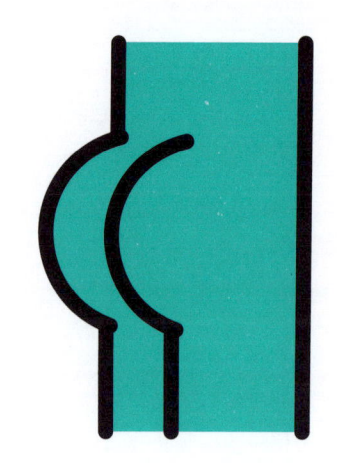

The butt of many a joke

Uranus gets a bum steer.

Yes, its poisonous gases mean no human can survive anywhere near it. And there's no denying a ring of debris surrounds the planet.

But if you're laughing right now, then according to scientists at the University of California, you're an idiot.

Grossly unfair, I say! Not to mention judgemental.

The planet is a cracker in its own right and has a great sense of humour. It — or those who find it funny — just don't deserve such a bum wrap. Sorry, there I go again. Bet you didn't know Uranus is around four times wider than planet Earth, orbits the Sun at a distance of 2.9 billion kilometres, and a single orbit takes around 84 years.

Still not impressed? Well, we're pretty proud of our solitary Moon, but Uranus has 27 of them. It also has those aforementioned rings — 13 to be precise. But there's absolutely no truth to the rumour there's a black hole nearby ...*

* Toilet humour, or more precisely 'scatological humour', encompasses all manner of jokes, puns and gags about pooing, weeing, farting, and other bodily functions. It's big with children and teenagers but most of us grow out of it by the time we reach adulthood. If you're looking for literary examples — low-brow deeds among high-brow reads — check out books like *Moby-Dick* by Herman Melville; *Ulysses* by James Joyce; *Don Quixote* by Miguel de Cervantes or, if you're after something more modern, *The Corrections* by Jonathan Franzen. And yes, one or two Uranus gags might have made it into my books, too.

Toasty

Ever wondered at what temperature bread becomes toast?

Well, at 154° Celsius, bread undergoes what's known as a 'Maillard reaction', so called after the French chemist, Louis-Camille Maillard, who first described it around 1910.

As the surface of the toast warms when heated by a 'dry' griller or toaster, new compounds are formed, the surface becomes brown and crunchy, and the toast develops an amazing richness and depth of flavour.

What's happening is the wheat proteins (like gluten) and added or naturally occurring sugars are reacting — and changing — because of the heat. It's an effect that just can't be produced by 'wet' cooking methods like steaming or braising.

Maillard reactions can begin at room temperature, but turning up the heat accelerates the process. It occurs with lots of other foods besides toast but be warned, the next stage of the cooking process — *pyrolysis* — means you've gone too far and your food will be burnt.

Blech

Foodies everywhere raced to Malmö, Sweden, last year to check out the unforgettably named Disgusting Food Museum.

The pop-up Museum featured 80 of the world's allegedly most 'disgusting' foods and even featured a restaurant open for lunch Monday to Friday. Unfortunately, none of the 80 disgusting foods was on the menu. This would have been annoying if you were hanging out for a Vegemite sandwich.

Say what? You heard right: alongside sheep eyeball juice, bull testicles, roasted guinea pigs, maggot-infested cheese, canned pork brains with milk gravy, frog smoothies, baby mice wine and fermented herring, the Museum also exhibited our very own Vegemite, musk sticks and witchetty grubs!

The Museum's creator Samuel West's aim was to explore the cultural subjectivity of food and challenge our ideas about why certain foods seem 'disgusting'. He did have the good grace to point out that he quite liked Vegemite ... but that it could be a very traumatic food experience for anyone who mistook it for Nutella!

Munira the Miraculous

It's rare for victims of long-term coma to regain consciousness and even if they do, recovery is never easy.

But tell that to Munira Abdulla. In 1991, at the age of 32, she was involved in a serious car accident while travelling with her son in the United Arab Emirates (UAE).

Her son emerged unscathed, but Munira suffered a severe brain injury. After being treated at various facilities in the UAE for several years, she was transferred to a hospital in London where she was declared to be in a vegetative state. She stayed in London for a few more years being fed by a drip and undergoing physiotherapy to make sure her muscles didn't waste away.

In 2017, she was then transferred to another hospital, this time in Germany.

And that's when it happened.

Just 12 months after arriving in Deutschland — and an incredible 27 years after her car accident — Munira regained consciousness. She has now returned to the UAE where she's undergoing more physiotherapy and rehab — but she's on her way to making a full recovery.

Bubble trouble

Ever seen a street performer blowing enormous, perfect bubbles?

Your kids lose their minds and beg you to repeat the magic at home. But when you try, you can only manage a half-hearted dribble.

Mathematics to the rescue!

What's critical is not the size of your bubble wand, or even the type of solution you're blowing. It's the wind speed.

Popular Science magazine has defined the ideal situation as follows:

$$U = \sqrt{\frac{5.6 \times \gamma}{\rho \times R}}$$

… where U is the wind speed, γ is the surface tension coefficient between the two fluids involved, ρ is the density of fluid you're blowing, and R is the radius of the bubble wand.

In other words, for a bubble wand with radius 1 centimetre, the optimal wind speed would be about 13 kilometres per hour.

So no excuses: get out there and dazzle some little 'uns with your bubble-blowing bravado!

The Blindfold King

It's hard to play chess. It's harder to play multiple games simultaneously. So how about trying it blindfolded?

Take a bow, Timur Gareyev, the 31-year-old grandmaster who played 48 games simultaneously while blindfolded in Las Vegas back in December 2016, smashing the previous world record. It took him 23 hours and he scored 80%.

The trick when playing multiple games while blindfolded is to keep the games distinct and separate. In Gareyev's case, his opponents had a mixed a variety of openings and Gareyev took black in half the games and played sharply and speedily. He also consulted mind experts beforehand and used a technique called 'memory palaces'.

Gareyev was born in Uzbekistan before moving to the US and is ranked inside the top 100 players in the world.

He has his sights now set on 55 simuls ... while blindfolded.

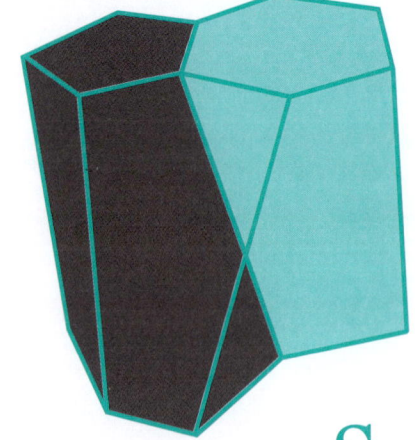

Scutoids?

Prism

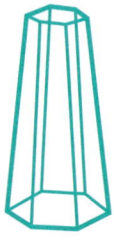

Frustum

Scutoid

Prismatoid

Last year we witnessed the discovery of an entirely new natural geometric shape. Don't know about you, but this was one of my personal highlights for 2018.

Why? Well, you and I are covered in them. They're called 'scutoids' and they're hiding in our skin. But before you run screaming to the mirror, don't worry, they're harmless ... and very small.

Up until recently, scientists didn't really understand the shape of our 'epithelial' cells. These are the building blocks of the structural tissue that form our skin.

But while using computer modelling to simulate the 3D packing of these cells, they got a surprise: a strange new shape appeared. It looked a bit like a prism, but while one end has 5 edges, the other has 6. This is down to the Y-shaped split dividing one of the prism's edges into two, creating a mini-triangle.

This 'twisted' prismatic shape had never been seen in nature before and didn't even have a name. The scientists chose 'scutoid' because it looks a bit like a part of an insect's thorax called the *scutellum*.

Welcome to the party, scutoid!

When is a sandwich ... not a sandwich?

Too late to ask the Earl o', I'm afraid. But thanks to *Atlantic Magazine,* we now have a rigorous 4-point test to determine what qualifies ... and what doesn't.

Their 'grand unified theory of the sandwich' proclaims that in order for a sandwich to call itself a sandwich, it must satisfy the following rules:

1. Structurally, the item must consist of two exterior pieces that are either separate or mostly separate;
2. Those pieces must be mainly carb-based (in other words, made of bread or bread-like products);
3. The item must have a mainly horizontal orientation (in other words, it has to sit flush with a plate, not perpendicular to it); and
4. The item must be portable.

Welcome to the sandwich club Oreos, burgers, *roujiamo** and ice-cream sandwiches.

Too bad for the wraps, tacos, open-faced sandwiches and hot dogs of the world. You don't make the cut.

*See page 160.

Say cheese!

It's fair to say you would have had to have been living under a rock, or in a hole (a black hole no less), not to have been utterly blown away by the sensational images revealed to the world in April 2019.

After decades of hard work, a team of international astronomers unveiled a picture of a massive gravity trap in the Messier 87 galaxy, some 55 million light years away …

In regular terms, THAT'S A PHOTO OF A BLACK HOLE! Or at least its silhouette.

But aren't black holes invisible? Don't they swallow light (and just about everything else)? And isn't 55 million light years quite a long way away? Valid and true questions, fellow space and time travellers. Capturing these images was nothing short of miraculous. And here's how they did it.

Back in April 2017, some of the world's most brilliant scientists linked 8 huge telescopes across the globe for the very first time. What was formed was a 'virtual telescope' with an aperture as big as the distance between the two farthest telescope stations — at the South Pole and in Spain — which meant it was nearly as big as the diameter of the Earth! They christened this optical orpheus the Event Horizon Telescope (EHT). The EHT is about 4000 times more powerful than the mighty Hubble and an extraordinary scientific achievement in and of itself.

But hold on, there's more. When it swung into action, the EHT used a bit of good old VLBI (I know you know that VLBI represents a technique called very-long-baseline interferometry but we have to bring everyone along with us here!) which allowed it to achieve an 'angular resolution' of 20 micro-arcseconds (again I know you know what 20 micro-arcseconds means; but to the uninformed that means it had enough sensitivity and resolution [that is, grunt] for you to be able to read a cryptic crossword in Hong Kong from your lounge room in Sydney). For more on the beauty of cryptic crosswords check out page 204, but let's not get distracted here ...

Collecting data from the 8 telescopes for 4 days gave the team an enormous amount of information to crunch — an estimated 5 petabytes (5000 terabytes). There was so much, it had to be physically transported on disk to a central location as current internet speeds simply couldn't handle the volume.

The data were synchronised and multiple calibration and imaging methods then kicked in, revealing a beautiful ring-like structure with a dark central region — the black hole's shadow. And what a sight it was. Scientists are now talking of time *before* the image was created and time *after*. It was that significant.

But why was it significant, Adam? Well, apart from the sheer scientific brilliance involved in obtaining the images, insights into these massive structures help us better understand physics and allow us to test observation methods and theories (like Einstein's theory of general relativity).* Black holes tend to deform spacetime and although Einstein's theory has been proven correct for smaller-mass objects like the Earth and the Sun, it hasn't been directly proven for black holes and other super-dense matter.

The EHT was also able to measure the radius of the black hole's 'event horizon' and determine its mass. We now believe it to be around 6.5 billion times the mass of the Sun. The EHT also showed that the black hole's silhouette was circular, more or less, which is great as most of us have grown accustomed to thinking of a hole as something round ... rather than square. And the term 'black box' has already been taken, anyway.

* According to EHT director Sheperd Doeleman from the Harvard-Smithsonian Center for Astrophysics, the images 'verified Einstein's theories of gravity in this most extreme laboratory'. It's not the first test that general relativity has passed — the theory has survived many challenges over the past 100 years — but it's a crucial one and once again testament to Einstein's towering genius.

The maths of art

If you read the intro to my last book (*Top 100*, thanks for asking, available where all good books are sold ...) you might recall me discussing just why I love numbers and mathematics so much.

I humbly submitted that maths was surely the most powerful tool we humans have to help us look around and understand, well, just what the heck is going on.

Now, it's not a competition, of course, but I also mentioned that I find the beauty in numbers outweighs even the breathtaking artistic achievements of, say Picasso with his (undeniably spectacular and revolutionary) *Les Demoiselles d'Avignon*. Big call. But of course it's just my humble opinion.

So you can imagine my joy when I came across the work of Dr Robert Fathauer, whose artistic statement says:

> Mathematical structure is evident throughout the natural world. My work incorporates the mathematics of symmetry, fractals, hyperbolic geometry and more, blending it with organic and inorganic forms found in nature. This synthesis allows me to create novel prints and ceramic sculptures that derive their appeal from juxtapositions such as complexity/order and movement/balance.

Like his hero, MC Escher, before him, Dr Robert began with tessellations. He then branched into fractals before exploring the beauty of fractal knots. The rest, as they say, is history. Grab yourself a cup of coffee and head over to his site (robertfathauer.com) and explore the wonderful art of maths.

Or should that be maths of art? Either way, I don't think you'll be disappointed.

Now, if this doesn't show you the beauty of maths, I'm not sure what will.

This work, which was constructed using the program KnotPlot, is based on the 3-crossing trefoil knot.

I've reprinted it here with the kind permission of Dr Robert Fathauer.

It has 9 crossings and 'the form of 3 orthogonal spirals that descend as the curvature decreases'. Gorgeous.

NUMBER LAND

YOU ARE HERE
365

ADAM SPENCER

366

Going out with a bang ...

Cruising around Twitter looking for mathematical breakthroughs ... as you do ...

Well, as *I* do anyway, I happened upon a retweet from Professor Michael Murray and Greg Egan who wrote, 'congratulations to Charlie Vane, who just announced a superpermutation for n = 7 of length 5907. A new record!' That number is replicated here on the left if you want to remember it for the next pub trivia night.

'What's a superpermutation?' you may well ask. Indeed I asked myself that very same question.

So I dived on in and checked it all out. It is fascinating stuff. And it would be such a waste for me to go do all that learning and not pass it on.

Well, fear not, dear reader ... let's go.

A word of warning — it's a bit of a rabbit hole, but hey, this is *Numberland*, after all.

To 'permute' something is to rearrange its parts. If you permute the letters of the word PAST you can get the words TAPS and PATS and other 'words' like ATSP, SPTA, and so on.

You should be able to see that the only permutations of the numbers 1 and 2 are the strings 12 and 21.

Can you think about how many permutations there should be of the numbers 1, 2 and 3 and can you list all of those permutations? Don't stress too much if you can't — I'll be giving you the answer over the page.

Given how awesome permutations already are, what could possibly be a *super*permutation?

Well, a superpermutation of n symbols is a string that contains somewhere within it all of the permutations of those n symbols. Huh? Let's go back to our buddies 1 and 2.

The easiest way to create a superpermutation of 1 and 2 is to take all two of the permutations of these numbers and glue them together.

So take 12 and 21 and form the obvious superpermutation 1221. This is a superpermutation because reading along it you can find both:

1221 and 12**21**

That is, you can find all of the permutations of the numbers 1 and 2.

But can you see how we can improve this superpermutation of length 4, removing a digit to create a superpermutation of length just 3?

That's right, drop one of the 2s and get the superpermutation 121 which contains both permutations:

121 and 1**21**

This is the shortest possible arrangement to contain both 12 and 21 and we call it the 'minimal superpermutation' of the two objects 1 and 2. On this page I will use the shorthand #n to denote a superpermutation of n objects. It doesn't have to be the shortest one; any superpermutation can bear the # symbol.

I asked you a few minutes ago to consider how many permutations there are of the string 1, 2 and 3. If you think about the process of creating any given string, you should see that you have 3 choices for the number with which you start. Once you've chosen a starting number, you've got two remaining numbers that could fill the second spot. And for each of the 3 possible starting numbers for a string, once you've allocated one of the two possible second elements of the string, you've only got one number left to round it out. So there are 3 × 2 × 1 = 6 possible permutations of the numbers 1, 2 and 3.

To save time writing out all those terms, we write 3 × 2 × 1 as 3!, pronounced '3 factorial'.

The 6 permutations of 1, 2 and 3 are 123, 132, 213, 231, 312, 321 and you may be able to spot that I've listed them in a 'logical' order so that I don't miss any of them.

Often when doing mathematics that involves counting or listing a number of items, a logical choice of order can really help. So again you could just glue these 6 permutations together to create the trivial 18-digit long #3:

123132213231312321 and then by simplifying the double two that happens in the 6th and 7th spot we get a 17-digit #3.

But there is a much shorter #3. In fact, it's just 9 digits long. If you're really feeling up for a challenge ... try to find it. If you're not really feeling up for it, don't stress, I'll be showing you later.

I won't ask you to try to find the 33-digit string which is the shortest #4 but you should be able to find all 4! = 24 permuations of 1, 2, 3 and 4 in the beautiful 33-digit string:

123412314231243121342132413214321

Now, are these minimal superpermutations related?

Could we, for example, use the shortest possible #3 to generate the minimal #4?

Well, indeedy do you can ... and here's the rule (which we maths nerds call an 'algorithm'). To generate a #n from a given #(n − 1) write out the permutations in the original superpermutation #(n − 1) in the order in which they appear.

Duplicate each of them, placing the new symbol n between the two copies.

Squeeze the result back together again, making use of all available overlaps.

For example, to get from $n = 2$ to $n = 3$, we write:

$$121$$

... and break it into the two permutations:

$$12 \mid 21$$

Then add the new number 3 into duplicates of each of the permutations:

$$12\ 3\ 12 \mid 21\ 3\ 21$$

Compress this into:

$$1231221321$$

and cancel one of the middle 2s to get:

$$123121321$$

This is the #3 of length 9 I asked you about earlier.
    Try this yourself to get from $n = 3$ to $n = 4$, that is, to use:

$$123121321$$

to generate 1234123142312431213421324132413214321.

You'll find the answer ... in about 5 lines time.
    You ready? Spoiler alert!

123121321
123 | 231 | 312 | 213 | 132 | 321
123 4 123 | 231 4 231 | 312 4 312 | 213 4 213 | 132 4 132 | 321 4 321
123 4 **123 231** 4 **231 312** 4 **312 213** 4 **213 132** 4 **132 321** 4 321

I've left some spaces so you can see where the simplification now happens. We have started with six 3-digit permutations each written out twice and thrown in six 4s. So the first #4 we have here is 6 × 3 + 6 × 3 + 6 = 42 terms long. But each of the terms that are in bold are examples of the one term repeating immediately in the chain. These 'doubles' collapse to just be 'singles', for example, 123231 becomes just 1231. Doing this we lose 9 terms and we are left with the 33-term minimal #4:

123412314231243121342132413214321.

Now, applying this algorithm again you get a 153-character minimal #5.

You can just trust me on this ... or gain my eternal admiration by crunching it out yourself.

Anyway, the #5 you get is ...

1234512341523412534123541231452314253142351423154231245312435124315243125431**2**1345213425134215342135421324513241532413524132541321453214352143251432154321

But this is where things start to get really fun. You see the #5 given above isn't the only #5 of length 153. One is closely related to this one. You get it by switching all of the 4s and 5s in the second half of the string (after the bold 2).

There are 6 other minimal #5s which don't seem to be related to the one generated by the algorithm and its close cousin. There could be some deeply hidden relationship we just haven't spotted yet, or perhaps they are truly different. Only time and a lot of computer grunt will tell.

I say grunt because these 6 other minimal #5s were generated by maths gun Benjamin Chaffin in 2014. He didn't just find 6 other #5s of length 153, he showed that this is the minimal possible length of any #5.

The way he did this was to look at what are called 'wasted characters'. Look back at our #3:

123121321

We start with 123 and by adding 1 we create a new permutation 231. By extending to 12312 we now have 312. But nothing we add as a stand alone 6th character can give us another permutation. We have to add 13 to give us the permutation 213. So in adding two characters we have added one more than the minimum number we could have added. This lengthens the final #3 and we can think of it as having 'wasted' a character.

So when exploring #5s we know the first 4 characters 1234 are necessarily wasted to get us started. And there are 5 × 4 × 3 × 2 × 1 = 120 different permutations of 12345 that have to occur in our #5, so a #5 of length 153 can only afford 153 − 120 − 4 = 29 other wasted characters. If we can construct one with only 28 other wasted characters we have found a #5 of length 152.

Chaffin showed to generate the 120 permutations of 12345, you have to waste at least 29 characters.

Now, just when you're thinking, 'Okay, Adam, you're right this *is* pretty interesting ...' well, hold onto your calculators, because once we start investigating #6s ... it gets *really* interesting.

So far the minimal #s we are finding follow a basic pattern. The minimal #n has length $n! + (n-1)! + (n-2)! + ... + 2! + 1!$.

You can see that the minimal #3 above has length 3! + 2! + 1! = 6 + 2 + 1 = 9 and that the #4 is 4! + 3! + 2! + 1! = 24 + 6 = 2 + 1 = 33 characters long and adding 5! = 120 to this we get 153, the length of our mimimal #5.

People had pretty much accepted that this formula would keep giving us the minimal length of every #n. So adding 6! = 6 × 5 × 4 × 3 × 2 × 1 = 720 to 153 you get 873 and indeed as the algorithm gives you a 873-length superpermutation of ABCDEF.

But it is *not* the shortest #6. What, Adam? Yep, you read me right. For *n* = 6 and higher, the algorithm does not give us the minimal length #*n*.

Another superpermutation superhero, Robin Houston, discovered this not long after Chaffin found the other six #5s we mentioned above.

Houston realised that searching for superpermutations can be thought of as taking a long walk around a path populated by permutations. Walking from ABC to BCA then to BAC, for example.

This is similar to another famous piece of mathematics called 'the travelling salesman problem' which asks how efficiently can a salesman or woman visit a series of cities, hitting every city at least once, but not wasting time.

Well, the problem is so famous there are computer models you can download to solve examples. Using one such program called LKH, the big Hou loaded up the graph for 6 symbols and headed off to bed. Lo and behold in the morning his laptop had found a string of numbers which looked like a recipe for a 872-character #6.

He reconstructed the superpermutation, checked that it contained all 6! = 720 permutations ... wow, that's a day you really need to be concentrating at the office ... and BAM he had found one. It turns out that the 872-digit chain* ... is a #6.

Now, in the same way we applied the algorithm to the minimal #3 to generate our minimal #4 earlier, we can now use this #6 rather than the 873 #6 generated by the minimal #5 as our starting point to generate #7s and #8s, and so on. So in this sense, Robin's discovery impacts our search for all longer minimal superpermutations.

In fact, there are lots of 872-length #6s. But we're not sure if 872 is the shortest possible length of a #6.

Okay Adam, surely we are done now. Ha!

In October 2018, there was another breakthrough. Robin Houston tweeted that there had actually been a step forward back in 2011 that had slipped by unnoticed. He

* which is ... 12345612345
162345126345123645132
645136245136425136452
136451234651234 15623
415263415236415234615
234165234125 63412536
412534612534162534126
534123564 12354612354
162354126354123654132
654312645316243 51624
315624316524316254316
245316425314 62531426
531425631425361425316
452314652314 56231452
631452361452316453216
453126435126 431526431
256432156423 15462315
426315423615423 16542
315642135642153624153
621453621543 62153462
135462134562 13465213
462513462153 64215634
216534216354 216345216
342516342156 43251643
256143256413256 43126
543216543261534261354
261345261342561342651
342615324651324653124
635124631524 63125463
215463251463 25146325
461325463124 56321456
324156324516324561324
563124653214 65324165
324615326415 326145326
154326514362514 365214
356214352614 352164352
146352143651243615243
612543612453612435612
436514235614 235164235
146235142635142365143
265413625413652413562
413526413524613524163
524136542136541 23, my curious friend!

informed the twitterverse (well, a small subset of Twitter at least, he doesn't quite have Beyoncé-like follower numbers):

A curious situation. The best known lower bound for the minimal length of superpermutations was proved by an anonymous user of a wiki mainly devoted to anime.

It turns out that someone had actually made a tremendous observation about minimal superpermutations, but instead of it being presented in an academic journal or at a maths conference, it was posted on 4chan and copied across to an anime blog. It all came about because fans of the show *The Melancholy of Haruhi Suzumiya* were wondering about watching the episodes of the show in every possible order. A question about superpermutations!

An anonymous poster suggested that to watch the 14-episode first season of *Haruhi*, in every possible order, you would have to watch at least 93,884,313,611 episodes back to back. They also gave a more general formula for the lengths of minimal #n. According to the anonymous poster's calculations, any #n must be at least $n! + (n-1)! + (n-2)! + n - 3$ characters long.

This work was verified and written up in an official paper called 'A lower bound on the length of the shortest superpattern' crediting authors 'Anonymous 4chan Poster, Robin Houston, Jay Pantone and Vince Vatter'. Surely this is the only time in the history of mathematics an anonymous 4chan poster has had top billing on a paper!

Then, another character emerges. Greg Egan, an Australian science fiction writer and full-on part-time mathematician (seriously, could this get any nerdier?), found a new method for calculating the length of possible #s that according to Robin Houston 'smashes it out of the park'.

Egan has shown that any minimal #n must be no longer than $n! + (n-1)! + (n-2)! + (n-3)! + n - 3$ characters.

So when trying to work out the length of any minimal superpermutation (MSP) it is trapped between:

$n! + (n - 1)! + (n - 2)! + n - 3 \leq \text{MSP} \leq n! + (n - 1)! + (n - 2)! + (n - 3)! + n - 3$.

For small vales of n these bounds don't really look that impressive. For #3 we get $9 \leq \text{MSP} \leq 9$; #4 $33 \leq \text{MSP} \leq 34$; and for the case of #5 $152 \leq \text{MSP} \leq 154$.

But for $n = 6$ these bounds on the minimal superpermutation are $867 \leq \text{MSP} \leq 873$ and as n gets large they are massive improvements on the old $n! + (n - 1)! + (n - 2)! + ... + 2! + 1!$ we had been working with.

The old algorithm would produce a #7 of length 5913 but our new bounds tell us that the minimal #7 lies between 5884 and 5908.

And the tweet that started all of this for me was Greg Egan commenting that Charlie Vane had found a #7 of length 5907. Well, Robin Houston and Greg have improved on this and found a 5906-length #7. The search continues.

Why does the algorithm hold up to #5 but fall apart for #6? There is something going on here, perhaps something deep and beautiful, but at the moment we just don't know!

If you're feeling enthused by this frenzy and would like to get involved, search 'Superpermutations: the maths problem solved by 4chan' and watch a great episode of *Standupmaths* on this very topic.

Final score

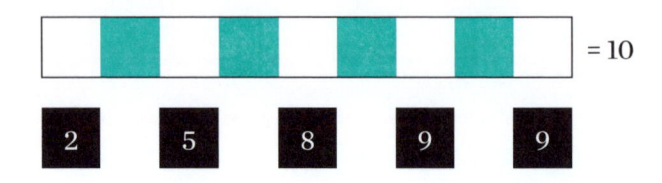

☐ 🟩 ☐ 🟩 ☐ 🟩 ☐ 🟩 ☐ = 10

2　**5**　**8**　**9**　**9**

☐ 🟩 ☐ 🟩 ☐ 🟩 ☐ 🟩 ☐ = 10

3　**3**　**5**　**8**　**9**

Reach the goal number by creating an equation using the provided numbers and your own choice of operations.

Numbers are placed in white squares, while operations (+, −, ×, ÷) are placed in green squares.

Order of operations matters here, per usual, and you can't use brackets!

Read the numbers off in order for your final score.

What is the equation that gives the highest score?

☐ 🟩 ☐ 🟩 ☐ 🟩 ☐ 🟩 ☐ = 10

2　**3**　**5**　**7**　**8**

Answers →

Page 7

Well, for a start, the answer is NOT 25.

Yes, the 3 cookies add up to 30 so they are worth 10 each. And the cookie plus two pairs of bananas is 14 so a pair of bananas is worth 2.

And the bananas plus two clocks adds to 8, so the clocks are worth 3 each.

So the trick is that the puzzle makes a lot of people race to thinking the 4th line is solved by reading clock (3) plus banana (2) plus banana (2) times cookie (10) which gives us 3 + 2 + 2 × 10 = 25. But no, no, no.

The bananas in equations 2 and 3 are banana pairs and have a value of 2. But in equation 4 we are only shown single bananas. Similarly, the cookie in equations 1 and 2 has 10 choc chips on it and has a value of, yep 10. But the cookie in equation 4 has only 7 chips! Uh oh ... now look at the clocks. The clocks that were set at 3 pm had a value of 3. What time is on the clock in equation 4?

So instead of 3 + 2 + 2 × 10 = 25 the 4th equation gives us 2 + 1 + 1 × 7 = 10. How'd you like those bananas?

Page 10

(a)

Page 26

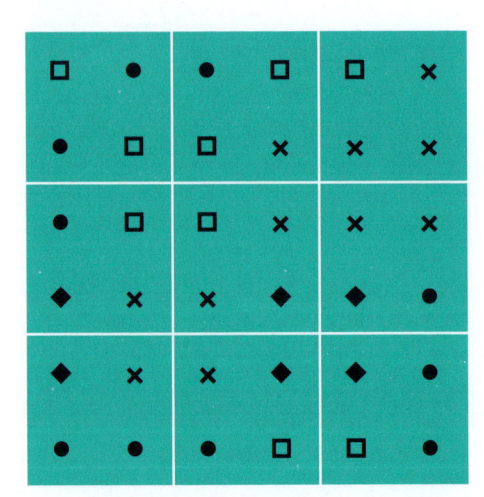

Page 35

90, 91, 92, 93, 94, 95, 96

Page 38

	3	+	5	−	1
	−	4	×	6	−
	1	×	3	×	6
	×	5	×	9	=
	1	×	2	=	1

	3	+	5	−	1
	−	4	×	6	−
	1	×	3	×	6
	×	5	×	9	=
	1	×	2	=	1

	3	+	5	−	1
	−	4	×	6	−
	1	×	3	×	6
	×	5	×	9	=
	1	×	2	=	1

Page 48

For the polygons, the green and white total areas are the same. This is actually a bit tricky to prove, particularly in the case where the number of sides is NOT a multiple of 4.

For the circle, if the points are equally spaced, the green and white total areas will be equal. Imagine the point inside the circle is not near the edge. Well, if you inscribe a regular polygon in the circle, we already know that the green and white components will match. The remaining area is a circular segment for each side, all of which have the same area (since the corresponding chords have the same length). If the point inside the circle lies outside this inscribed polygon ... it's been a long book, how about you just take my word for it :)

Page 57

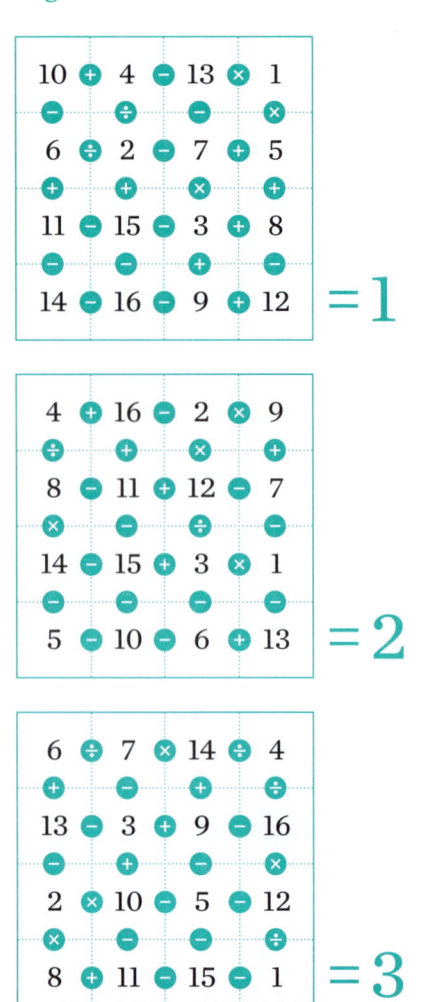

```
10 + 4 − 13 × 1
−    ÷    −    ×
6  ÷ 2 − 7  + 5
+    +    ×    +
11 − 15 − 3  + 8
−    −    +    −
14 − 16 − 9  + 12
```
$= 1$

```
4  + 16 − 2  × 9
÷    +    ×    +
8  − 11 + 12 − 7
×    −    ÷    −
14 − 15 + 3  × 1
−    −    −    −
5  − 10 − 6  + 13
```
$= 2$

```
6  ÷ 7  × 14 ÷ 4
+    −    +    ÷
13 − 3  + 9  − 16
−    +    −    ×
2  × 10 − 5  − 12
×    −    −    ÷
8  + 11 − 15 − 1
```
$= 3$

Page 72

Yes, there is. Do the following — outdoors and *carefully*! Stick the match in the cork and place it in the middle of the plate. Light the match and place the upturned glass over the top of it.

The match will heat the air in the glass which creates pressure against the water and allows some of the air to seep out. When the match finally goes out, the air starts to cool again and the ensuing relief in pressure will suction water up into the glass. Nifty!

Page 89

| 7 | − | 6 | × | 2 | + | 5 | + | 1 | = 1 |

76251

| 8 | × | 6 | − | 8 | × | 5 | − | 7 | = 1 |

86857

| 8 | ÷ | 3 | − | 3 | ÷ | 9 | × | 5 | = 1 |

83395

Page 105

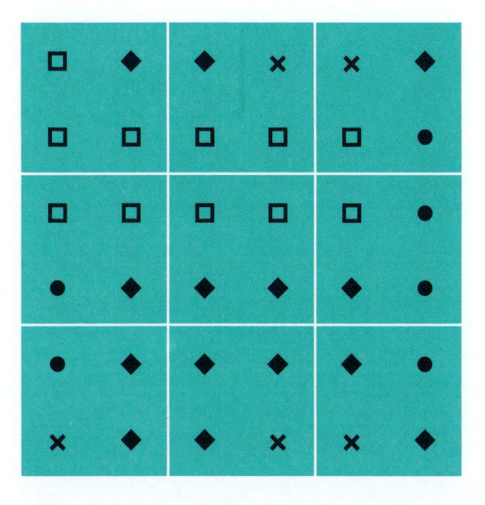

Page 112

You can generate p/q from the 'naive addition' p/q = (a + c) / (b + d)

Pages 114–115

(continued)

Page 126

Jim went on to thank the many repliers on Twitter for their birthday wishes with the following: 'As many people worked out, I am 70 = 2 × 5 × 7 (66 = 2 × 3 × 11), the smallest solution. Nobody suggested the next smallest, 114 = 2 × 3 × 19 (110 = 2 × 5 × 11). I can't be that old, as I am not on the list of the world's oldest people.'

He goes on to say that the easiest way to see that 70 is the smallest answer is 'it must be even, because the second smallest odd pqr is 3 × 5 × 11 which is too big. So $2pq = 2rs + 4$, or $pq = rs + 2$, with p, q, r and s all different odd primes. The smallest $\{p, q, r, s\}$ is $\{3, 5, 7, 11\}$, and lo that works: 5 × 7 = 3 × 11 + 2.

And as for me? I was 49 (7 × 7) and now I am 50 (2 × 5 × 5). This also happened when I turned 2, though at the time I didn't appreciate the fact.

The next few numbers for which this is true are:

1681, 1682 (41^2, $2 × 29^2$)

57,121; 57,122 (239^2, $2 × 169^2$)

1,940,449; 1,940,450 (1393^2, $2 × 985^2$)

Page 129

Unfortunately for our jumpy friend, despite her impressive efforts, she will never make it to the metre mark, since she keeps adding distance to smaller and smaller amounts.

Look, I didn't say it was a very interesting sport to watch ...

Page 131

Line 6, 7th from the end.

Page 134

I'm in second place! I passed the person in second place, after all — *not* the person in first place. Gotcha!

Page 138

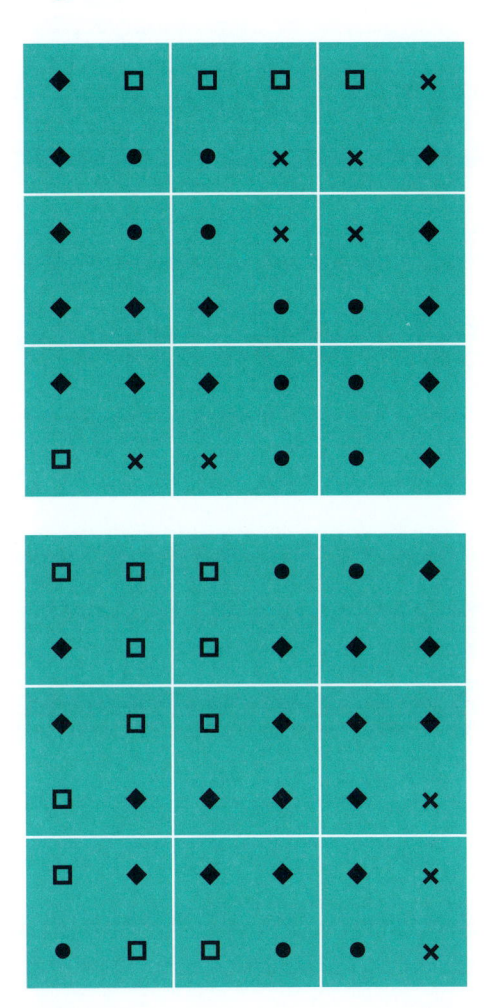

Page 141

Admit it — you said June, at least in the beginning, right? Read the question again and you'll see that the answer's right there in front of you. That's right, the third child's name is ... David.

Our brains tend to like to skip along and 'skim' information like this, which is why we so readily jump to the conclusion about the name. There's a good reminder here that it's worth paying a little bit of extra attention.

Page 142

Each page fits 10^6 zeroes. So we need $10^{100}/10^6 = 10^{94}$ pages.

Page 158

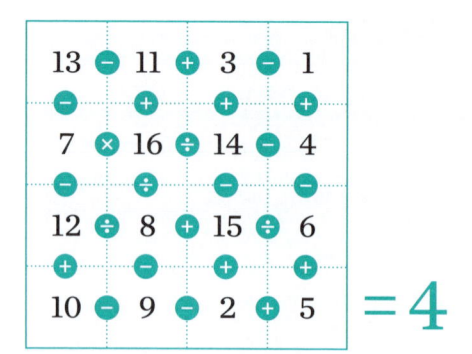

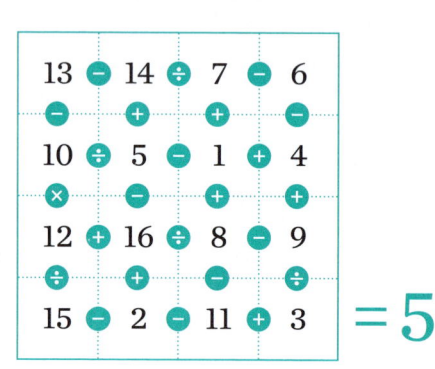

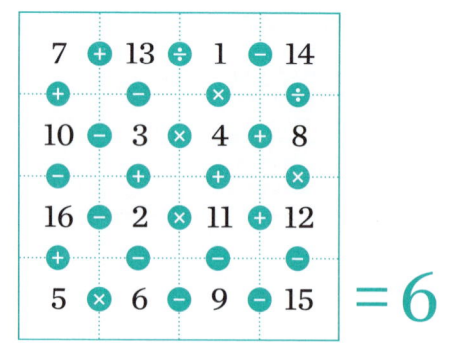

$$13 - 11 + 3 - 1$$
$$- \quad + \quad + \quad +$$
$$7 \times 16 \div 14 - 4$$
$$- \quad \div \quad - \quad -$$
$$12 \div 8 + 15 \div 6$$
$$+ \quad - \quad + \quad +$$
$$10 - 9 - 2 + 5$$

$= 4$

$$13 - 14 \div 7 - 6$$
$$- \quad + \quad + \quad -$$
$$10 \div 5 - 1 + 4$$
$$\times \quad - \quad + \quad +$$
$$12 + 16 \div 8 - 9$$
$$\div \quad + \quad - \quad \div$$
$$15 - 2 - 11 + 3$$

$= 5$

$$7 + 13 \div 1 - 14$$
$$+ \quad - \quad \times \quad \div$$
$$10 - 3 \times 4 + 8$$
$$- \quad + \quad + \quad \times$$
$$16 - 2 \times 11 + 12$$
$$+ \quad - \quad - \quad -$$
$$5 \times 6 - 9 - 15$$

$= 6$

Page 159

There is actually a simple solution to the riddle. You must ask one guard which door the *other* guard would say leads to the castle. It doesn't matter which guard you ask, as both guards will indicate the same door, which must be the door which *doesn't* lead to the castle.

Let's go a little deeper. Call the safe door 'A', and the certain-death door 'B'. Presume the guard you ask is the one who always lies. If this is the case, then he will obviously try to lead you astray since he will falsely claim that the other guard, who must tell the truth, will point you to B. So, door A is safe.

Now, presume that the guard you asked always tells the truth. In this scenario, he will tell you, honestly, that his counterpart (the liar) will point you towards door B. So, door A must be the safe bet.

In both scenarios, the outcome is identical. Door A it is! Although that doesn't mean our hero in *Labyrinth* is completely out of danger. You'll just have to watch the film to see what I mean ;-)

Page 167

If B is a knight, then he is telling the truth, so A is also a knight. But A said B is a knave, which is a contradiction. So B must be a knave, which means A is telling the truth and therefore A is a knight.

Page 169

| 7 | − | 8 | − | 8 | + | 6 | + | 5 | = 2 |

78865

| 8 | × | 5 | − | 7 | × | 6 | + | 4 | = 2 |

85764

| 9 | ÷ | 6 | ÷ | 6 | + | 7 | ÷ | 4 | = 2 |

96674

Page 182

Page 196

Atticus Finch (from Harper Lee's classic novel *To Kill a Mockingbird*).

Page 197

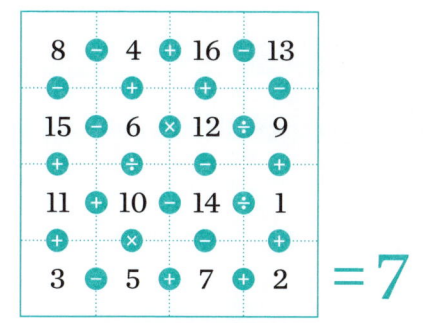

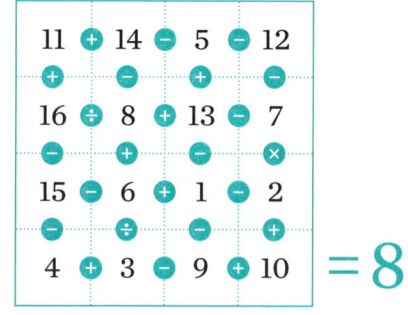

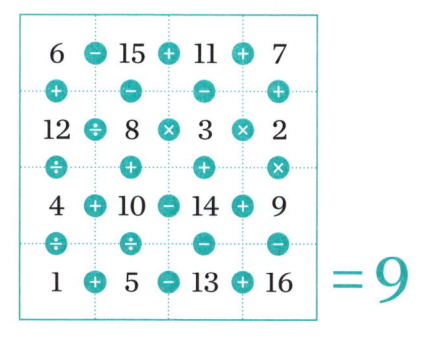

$$8 - 4 + 16 - 13$$
$$-\quad+\quad+\quad-$$
$$15 - 6 \times 12 \div 9$$
$$+\quad\div\quad-\quad+$$
$$11 + 10 - 14 \div 1$$
$$+\quad\times\quad-\quad+$$
$$3 - 5 + 7 + 2 = 7$$

$$11 + 14 - 5 - 12$$
$$+\quad-\quad+\quad-$$
$$16 \div 8 + 13 - 7$$
$$-\quad+\quad-\quad\times$$
$$15 - 6 + 1 - 2$$
$$-\quad\div\quad-\quad+$$
$$4 + 3 - 9 + 10 = 8$$

$$6 - 15 + 11 + 7$$
$$+\quad-\quad-\quad+$$
$$12 \div 8 \times 3 \times 2$$
$$\div\quad+\quad+\quad\times$$
$$4 + 10 - 14 + 9$$
$$\div\quad\div\quad-\quad-$$
$$1 + 5 - 13 + 16 = 9$$

Page 201

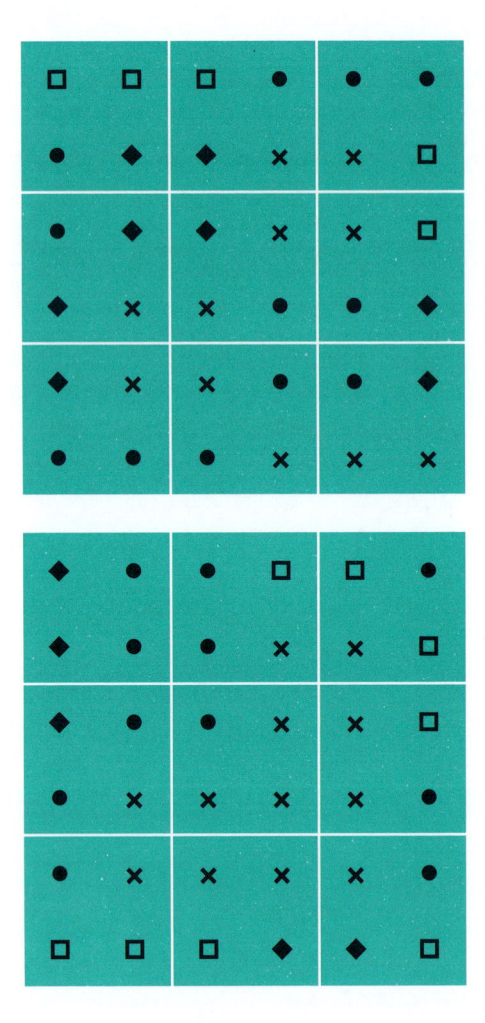

Pages 204—205

One-third of twelve still an odd number

The answer to this is gorgeous. It throws you off the scent because you probably start by thinking that one third of twelve is four. You'd be correct, but that's no help here.

Instead, note that the word 'twelve' has 6 letters. So one third of that word is the pair of letters EL. And another word for 'still' is EVEN; 'still if he does it …' means the same as 'even if he does it'. Putting those together, EL and EVEN give us ELEVEN which is 'an odd number'.

Hey, I said 'gorgeous', not 'easy'.

Unusually fab, iconic name for a series

Fibonacci.

Tahitian crime out of control — one can control figures

Arithmatician.

Pages 207–208

Here are the answers, straight from the great man, Iain Johnstone, himself:

ONE — Harry styles old direction found in the money

Let's start with a gimme. WTMI in this clue!

Pretty straightforward pointer to the group which formerly featured Harry Styles: ONE Direction.

The signposts found and in (the) direct us to look for our solution hidden or contained in another part of the clue.

In this case mONEy.

Couple of other things going on: Styles is used as a verb but also forms the proper noun we need, while direction seems to be pointing somewhere, but is also a part of ONE old thing Harry was involved with.

TWO — Take the first tune we offer otherwise you can't tango

Signifiers such as first, open, start, etc, will often tell you that initial letters from words in the clue are needed to spell out the solution.

So here we find Tune We Offer spells out TWO, which everybody knows is the number it takes to tango.

THREE — How many stooges are over there?

Anagrams are a common and potent weapon in the cryptocruciverbalist's armoury.

Look for words which invoke movement, change, action, etc.

Here's a quick and easy one.

The word over is one of many which might tell you that somewhere else in the clue you'll find a word (or words) which you'll need to rearrange to reach the solution word.

Flip over the letters of THERE and you should find the number of stooges you're looking for.

[This is a clue where there's a cultural reference which relies upon a bit of assumed knowledge. There are plenty of things that come in rhetorical/numerical groups but here we are dealing with a comedy team of which there are famously THREE.]

FOUR — Golfer's call to the boundary

This one's a neat homophone, indicated by call which is a signifier pointing us to audible clues. These may be part of an answer — such as a syllable — but in most cases, including this one, are the whole word we are looking for.

There's a bonus clue for sports fans in the references to golf and cricket. Assumed knowledge is that a boundary in cricket is worth four runs, and that the golfer's warning cry of 'fore', which means 'look out up ahead', is a homonym for our answer.

FIVE — Following four in the iron

Conventional symbols are involved here.

Roman numerals and chemical symbols are always worth bearing in mind.

Whenever a clue specifies a number it's worth seeing if the corresponding letters of the Roman numerical system might be involved in the solution.

Here we're looking for something involving four.

We know that four in Roman numerals is IV.

We also know that the chemical symbol for iron on the periodic table is FE.

So when we put IV in FE we get FIVE, which follows four!

SIX — Line of symmetry lost a half dozen

This one's just unpicking the wordplay in the first part of the clue to arrive at the second part which is a straightforward Quick Crossword clue for the same solution.

A line of symmetry is an axis.

Symmetry also tells us something might be reflected, or rendered backwards, and axis backwards is sixa but we are told the a is lost.

This leaves us with SIX, which is a half dozen.

SEVEN — First woman runs up between poles to find enough brides for the brothers

References to the Bible and other classical mythologies and canonical literature as well as popular culture are common touchstones in cryptic puzzles.

In this instance we're asked to put together the oldest bit of the Old Testament and a 1950s Hollywood musical.

The Bible's first woman is EVE.

Poles are North and South, but here we are asked to go up from South to North.

EVE moves from South to North. SEVEN.

If Brides for Brothers are not your thing feel free to use any combination of magnificent Dwarfs or Samurai at your discretion.

EIGHT — Rowing crew noisily consumed inbound freighter

A rowing crew is an EIGHT, no tricks yet and a common enough clue used in Quick Crosswords.

The cryptic elements come next.

Noisily tells us to look for an audible clue or homophone for consumed, which is ate, a homophone of EIGHT.

And just for good measure both in and bound can indicate that our answer is contained in another part of the clue, in this case the word frEIGHTer.

NINE — Trafalgar, Tiananmen and Spongebob's pants need a good day's meals to start working towards five

Some lateral thinking here.

Trafalgar, Tiananmen and Spongebob's pants: Square, square, square. Three square is NINE.

'Three square' is also a euphemism for breakfast, lunch and dinner, the standard for meals in a day.

And the working day is often referred to as NINE to five.

So just some unusual and circuitous ways to arrive at the number NINE.

TEN — Half your score evenly strewn across net return

A score is twenty, and half of that is TEN.

But wait there's more.

When odd or even qualities are invoked you should look for the odd or even letters in a word, in this case the even letters of sTrEwN are also TEN.

And if you return net, i.e. read it backwards, you get ... TEN!

Page 225

| 9 | − | 9 | ÷ | 3 | − | 8 | + | 5 | = 3 |

99385

| 9 | − | 6 | ÷ | 3 | − | 5 | + | 1 | = 3 |

96351

| 9 | ÷ | 6 | × | 4 | − | 7 | + | 4 | = 3 |

96474

Page 235

If Barrington is a knight, then he's telling us Adam is a liar, so Adam must be a knave.

If Barrington is a knave, then he's lying when he says Adam would call him a knave — that is, Adam is a liar, so Adam must be a knave again.

So Adam is a knave regardless of what Barrington is.

That means that everything Adam says is false, so Spencer is a knight.

Since Spencer is a knight, he's telling the truth when he says both Adam and Barrington are knaves.

So, friends, Spencer is a knight. The other two, knaves.

Page 238

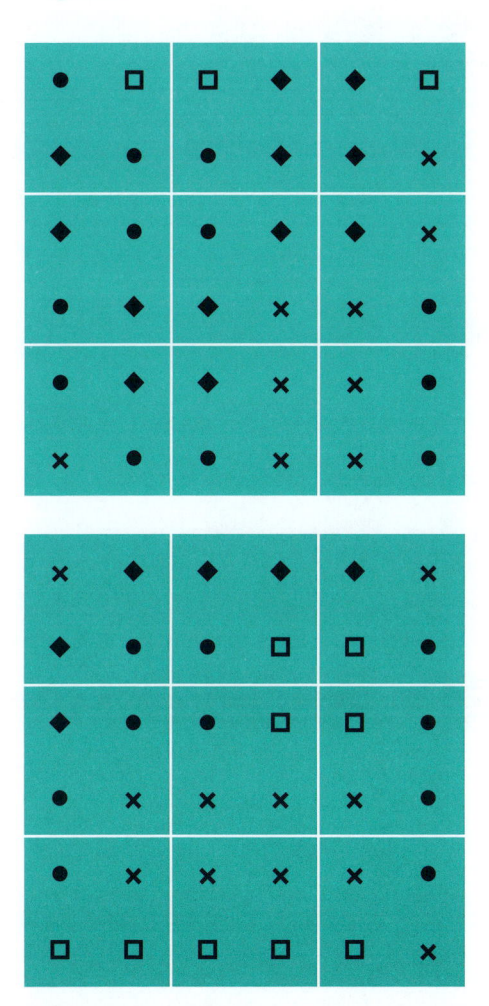

Page 241

| 7 | – | 8 | + | 7 | – | 1 | – | 1 | = 4 |

78711

| 9 | × | 5 | – | 8 | × | 6 | + | 7 | = 4 |

95867

| 8 | × | 5 | ÷ | 4 | − | 7 | + | 1 | = 4

85471

Page 255

| 9 | − | 6 | − | 6 | ÷ | 3 | + | 4 | = 5

96634

| 6 | + | 1 | ÷ | 7 | − | 8 | ÷ | 7 | = 5

61787

| 9 | × | 3 | − | 7 | × | 4 | + | 6 | = 5

93746

Page 258

Here's the solution, from my man the Surfing Scientist himself:

'Flip both timers over. When the 4-min timer is empty, flip it over (the 7-min timer has 3 minutes left in the top). When the 7-min timer is empty, flip it over (there is now 1 minute left in the top of the 4-min timer).

When the 4-min timer is empty for the second time, the 7-min timer has 1 minute worth of sand in the bottom half. Flip it over so there is now 1 minute in the top. When the 7-min timer empties, 9 minutes have elapsed.'

Page 264

There are a few ways to skin this particular cat ...

- G is a two-digit sixth power, i.e. 64.
- B and F are two-digit primes with the same last digit and sum 54 — they must be 17 and 37.
- If B = 17, then E is a three-digit cube starting with 1, i.e. E is 125. Then D must be a three-digit square starting with 5 and ending with

- 4 — impossible. Thus B = 37 and F = 17.
- E is a three-digit cube starting with 3 — i.e. E is 343.
- D is a three-digit square starting with 3 and ending with 4 — i.e. D is 324.
- Since C is a two-digit number and H = C + 100, H starts with a 1.
- Thus A is a three-digit multiple of 11 starting and ending with 1 — i.e. A is 121.
- Since C starts with 6, the second digit of H is 6.
- Note that E and G are cubes, and A, D, and G are squares. Thus C and H are not squares or cubes.
- The last digit of C and H can be any of (0, 1, 2, 3, 5, 6, 7, 8).

Note that if you allow there to be four different squares on the board (instead of three), there is a unique solution — H must be 169 (and C is 69).

Page 274

Working backwards:
- There is a unique 3.
- There are six different choices for 2.
- For each 2, there are three different choices for 1.

Thus the total number of paths is 1 × 6 × 3 = 18.

Page 279

- If Essendon is the footy player, then both Collingwood and Sydney are telling the truth, leaving no possible knave. Thus Essendon can't be the footy player.
- Thus both Collingwood and Sydney are lying, leaving Essendon as the knight.
- Since Essendon tells the truth, Sydney is the footy player, leaving Collingwood as the knave.

Page 283

My path (notation is (a, b) to represent the square in row a and column b):
- Start at (1, 1)
- Move right one square to (1, 2)
- Move down two squares to (3, 2)

- Move right three squares to (3, 5)
- Move left four squares to (3, 1)
- Move down one square to (4, 1)
- Move down two squares to (6, 1)
- Move right three squares to (6, 4)
- Move up four squares to (2, 4)
- Move right one square to (2, 5)
- Move left two squares to (2, 3)
- Move down three squares to (5, 3)
- Move up four squares to (1, 3)
- Move right one square to (1, 4)
- Move right two squares to (1, 6)
- Move left three squares to (1, 3)
- Move down four squares to (5, 3)
- Move right one square to (5, 4)
- Move right two squares to (5, 6)
- Move up three squares to (2, 6)
- Move left four squares to (2, 2)
- Move left one square to (2, 1) (Finish)

Page 286

| 9 | × | 2 | − | 7 | − | 4 | − | 1 | = 6 |

92741

| 9 | + | 9 | × | 3 | − | 6 | × | 5 | = 6 |

99365

| 9 | ÷ | 6 | × | 5 | × | 2 | − | 9 | = 6 |

96529

Page 292

| 9 | × | 2 | − | 9 | − | 7 | + | 5 | = 7 |

92975

| 6 | + | 4 | ÷ | 5 | + | 1 | ÷ | 5 | = 7 |

64515

| 8 | ÷ | 7 | × | 6 | + | 1 | ÷ | 7 | = 7 |

87617

Page 293

Noting that the 4 is unique, let's look at the 2s.
- The CORNER 2s have three choices for 1, and one choice for 3. There are six corner 2s, giving 6 × 3 × 1 = 18 paths in this case.
- The EDGE 2s have two choices for 1, and two choices for 3. There are six edge 2s, giving 6 × 2 × 2 = 24 paths in this case.

Thus, there are 18 + 24 = 42 different paths.

Page 303

| 9 | + | 8 | − | 7 | × | 2 | + | 5 | = 8 |

98725

| 7 | − | 8 | ÷ | 8 | + | 6 | ÷ | 3 | = 8 |

78863

| 8 | × | 7 | ÷ | 4 | − | 6 | × | 1 | = 8

87461

Page 306
- FORTY
- ONE
- THOUSAND
- FOUR
- 'eleven plus two'

Page 307
The answer is 2 — it's the number of 'loops' or circles formed by the shape of the numbers :-)

Page 314
Let's once again label the guards — A, B and C.
- Ask guard A: is guard B more likely to tell the truth than guard C?
- If guard A always tells the truth, then 'yes' implies that guard C is always lying, while 'no' implies that guard B is always lying.
- If guard A always lies, then 'yes' implies that guard C is always truthful, while 'no' implies that guard B is always truthful.
- Thus 'yes' indicates that guard C is not random, while 'no' indicates that guard B is not random (note that if guard A is random, this still works because guards B and C are both not random regardless of A's answer).
- Ask the non-random guard: if I asked you 'does the first route lead to treasure?', would you answer 'yes'?
- If the first route does lead to treasure, the truth-teller will say yes, while the liar will lie about saying no (i.e. they will also say yes).
- If the first route leads to no treasure, the truth-teller will say no, while the liar will lie about saying yes (i.e. they will also say no).
- Thus, if they say yes, pick the first route. If they say no, pick the second route.

400

Page 318

| 7 | − | 8 | − | 5 | + | 5 | × | 3 | = 9 |

78553

| 9 | × | 3 | − | 7 | × | 2 | − | 4 | = 9 |

93724

| 9 | × | 8 | ÷ | 3 | − | 5 | × | 3 | = 9 |

98353

Page 322

All sets of inequalities have at least one solution.

The most is for the inequality $a < b > c < d > e < f$ and its reverse $a > b < c > d < e > f$, with each having 61 solutions.

Page 326

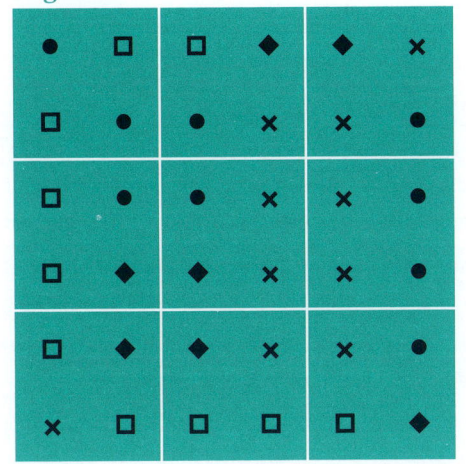

Page 329

Left to right, top to bottom:

14, 5, 11, 21, 28, 22, 8, 9, 15, 17, 6, 24, 25, 23, 12.

Clockwise from top left:

Inner pentagon: 28, 22, 15, 6, 9.

Middle pentagon: 5, 11, 24, 23, 17.

Outer pentagon: 14, 8, 12, 25, 21.

Page 340

The strategy is to start at 0 and work up.

- Clearly the smallest 4-term sequence is: 3, 2, 1, 0
- To get a 5-term sequence, the second term needs to be at least 3, thus the first term needs to be at least 3 more than the largest square less than or equal to it. Suppose the first term is between n^2 and $(n + 1)^2$. Then n^2 and $(n + 1)^2$ must differ by at least 4, i.e. $2n + 1 \geq 4$, or $n \geq 2$. Thus the first term is $2^2 + 3 = 7$, giving the sequence: 7, 3, 2, 1, 0
- The 6-term sequence follows a similar approach — the first term must be at least 7 more than the largest square less than or equal to it, that is $(n + 1)^2 \geq n^2 + 8$, thus $2n+1 \geq 8$ or $n \geq 4$. Our first term is then $4^2 + 7 = 23$, and the sequence is: 23, 7, 3, 2, 1, 0

- Repeating the pattern, for our 7-term sequence we have $2n + 1 \geq 24$, or $n \geq 12$. Thus the first term is $12^2 + 23 = 167$, and our sequence is: 167, 23, 7, 3, 2, 1, 0
- Finally, for the 8-term sequence we have $2n + 1 \geq 168$, or $n \geq 84$. Thus our first term is $84^2 + 167 = 7223$, and the sequence is: 7223, 167, 23, 7, 3, 2, 1, 0.

Page 344

I found two solutions: 2 cabins and 8 cabins. Note that the 2 cabin solution consists of one red and one green cabin, so this can be excluded if we assume that there must be red, green and blue cabins.

My mate Sean tested all possibilities up to 68 cabins and found no further solutions. Actually, the S-Man went a step further and wrote a program to confirm 2 and 8 are the only solutions less than 100,000.

There isn't much of a pattern to where the process fails for each number >8, but here are some obvious facts: there must be an even number of cabins since an odd number would mean no cabin is opposite another.

Also, the number of cabins must have a remainder of 0 or 2 when divided by 6, because if it has a remainder of 4 then there will be an odd number of blue cabins, which means at least one blue cabin is not opposite a blue cabin.

Another mathsy mate, Gareth, tells me that 'unfortunately while I am 99.999% sure it is the case, I have not been able to prove that 8 is the only possible solution'. In all fairness, that's likely because of my howling print deadline. Nuff said, G-Man.

Page 376

| 9 | ÷ | 9 | + | 8 | ÷ | 2 | + | 5 | = 10

99825

| 9 | + | 8 | ÷ | 3 | – | 5 | ÷ | 3 | = 10

98353

| 8 | × | 5 | ÷ | 2 | – | 7 | – | 3 | = 10

85273

See you next time!

... and online, of course —
adamspencer.com.au